Analysis of Science, Technology, and Innovation in Emerging Economies

Clara Inés Pardo Martínez
Alexander Cotte Poveda
Sylvia Patricia Fletscher Moreno
Editors

Analysis of Science, Technology, and Innovation in Emerging Economies

palgrave
macmillan

Editors
Clara Inés Pardo Martínez
School of Management
Universidad del Rosario
Bogotá, Colombia

Sylvia Patricia Fletscher Moreno
Colombian Observatory of Science
and Technology (OCyT)
Bogotá, Colombia

Alexander Cotte Poveda
Universidad Santo Tomás
Bogotá, Colombia

Colombian Observatory of Science
and Technology (OCyT)
Bogotá, Colombia

ISBN 978-3-030-13577-5 ISBN 978-3-030-13578-2 (eBook)
https://doi.org/10.1007/978-3-030-13578-2

This Palgrave Macmillan imprint is published by the registered company Springer Nature Switzerland AG.
The registered company address is: Gewerbestrasse 11, 6330 Cham, Switzerland

Foreword

Research and development (R&D) occupies a central place in all schools of economic thought, including the emphasis of several of them on the innovation systems that make the link between R&D and economic activity. Strong education and labor training systems are also essential to provide the adequate high-quality labor required for all these activities and have also been the focus of significant attention. All of them are required to increase productivity, economic growth, and the standard of living of the population. Advance in all these areas, but particularly in R&D, is seen as essential to reduce the gap that separates emerging and developing countries from more advanced nations. Due to these links, economic theory recognizes the importance of developing strong science and technology as well as equally robust production sector policies, that is, "industrial" policies, though not in the narrow sense of manufacturing. This is both due to the incapacity of the innovator (firm, group of researchers, or individual persons) to fully appropriate the benefits of their innovations and due to the strong interactions among innovators and firms that are required to develop new economic activities—"externalities", in the economic jargon, and scale economies of different nature.

The "structuralist" tradition in economic thought emphasizes, in turn, that the diversification of economic structures toward activities with greater technological contents is essential and indeed the basic source of economic growth. This tradition includes remarkable figures and intellectual traditions, such as Joseph Schumpeter, Hollis Chenery—the first World Bank chief economist—Raúl Prebisch, and the Latin American structuralist and neo-structuralist traditions, the evolutionary school of economics associated with the work of Richard Nelson, among others, and the "New Structuralist Economics" of a more recent World Bank chief economist from China, Justin Yifu Lin. The incapacity to develop these new higher-technology activities is seen in the recent literature as one of the factors that may generate a "middle-income trap": a tendency to stagnate at middle-income levels, a phenomenon in which most Latin American countries are probably the best examples in the world.

This is linked to the inadequate levels of R&D spending and the relatively low share of private sector spending in this area, which are remarkable features in Latin America. According to the most recent World Bank and UNESCO data, Latin America and the Caribbean spent about 0.9% of GDP in R&D, well below the level of upper middle-income countries (1.7% of GDP) and those of developing countries in the East Asia and the Pacific and OECD members (about 2.5% in both cases). The Latin American average is associated with the relatively higher levels of spending by Brazil (1.2%), which followed by a large margin by Argentina, Costa Rica, and Mexico (in the 0.5–0.6% range) and, even more, Chile, Colombia, Cuba, Ecuador, and Uruguay (in the 0.3–0.4% range); the rest of the countries spent minimal amounts (about 0.1% of GDP or less) or lack information (Venezuela). On the positive side, the trend has been positive for most countries, with the major exceptions of Cuba and Panama.

When we add the resources invested in the broader category of "innovation", the results are somewhat more positive. For example, in the excellent publication by the Colombian Observatory of Science and Technology, led by one of the co-editors of this volume, Clara Inés Pardo Martínez, which uses the most recent OECD guidelines to measure spending in R&D and innovation, estimates that the broader range of outlays in science, technology, and innovation reaches close to 0.7% of GDP in 2017 versus less than 0.3% for R&D. There are also somewhat good signals in this area, including rising private spending, access to information and communications technologies, and publications in indexed scientific journals. On the negative side, it shows the deplorable levels of patenting innovations by Colombian researchers and firms, and the rather negative trend for public-sector spending and the university system. Some of these results are presented in Chap. 2 in this volume.

The link between inadequate levels of R&D with the "middle-income trap" faced by Latin America seems to be clear. The major reflection is the fairly broad de-industrialization trend that the region has been facing, which is furthermore "premature" according to the existing literature, as it has happened at much lower levels of per capita GDP than when it took place in the now industrialized countries. This is, of course, in open contrast to the experience of East Asia, where industrialization and technological deepening has continued, not only in China but also in several emerging economies in that region, some of which have made the transit to high-income ones, notably the Republic of Korea. This is accompanied by high levels of R&D spending, which has boomed in China, reaching levels of more than 2% of GDP in recent years, and making the Republic of Korea one of the outstanding performers in this area, with spending that surpasses now 4% of GDP.

There are, of course, interesting cases in Latin America in terms of R&D, which have had positive effects on the development of specific sectors. Perhaps the more remarkable are in agricultural research, which has been reflected in the leadership that some countries have in agricultural production and exports at the world level, such as soybeans and maize from Brazil and Argentina, which has had positive externalities on smaller countries (Bolivia and Paraguay). There are a few other examples, such as the aircraft industry and the deep-oil developments in Brazil. But these are not generalized trends and are highly concentrated in Brazil, the Latin American leader in R&D, as we have seen.

Doing research on these issues in relation to the emerging economies is, therefore, essential to bring attention and upgrade policies in this area. It is important, in particular, that research be close to the policymakers themselves, to have a stronger policy impact. This book is a remarkable case in this sense. It is edited by the two Colombian professors, including the head of the Observatory of Science and Technology, and includes contributions from several other Colombian professors and their partners in Ecuador, Mexico, and Spain.

It covers a diverse set of interrelated issues. It includes broad policy issues, such as the influence of science and technology in the development of three major Latin American cities, the role of open access policies in science in Colombia, technological policies and gender inclusion in science and technology in Ecuador, and the broad effects of R&D on technical change in Latin America and the Caribbean. It also deals with sectoral topics: biotechnology and pharmaceutical development, as well as water access in Colombia, and the aerospace cluster in Querétaro, Mexico, and what it can learn from a similar cluster in Madrid. And it includes analysis of trends in scientific education, R&D, patenting, and the national perceptions of science, technology, and innovations in Colombia. The country most included in the analysis is Colombia, but also Ecuador, Mexico, Chile, and Latin America and the Caribbean in general.

I am grateful to the editors of this volume, Alexander Cotte and Clara Inés Pardo Martínez for inviting me to write the preface for this very interesting and novel research volume, and congratulate Palgrave Macmillan for an excellent addition to their outstanding book collection. I hope this research will contribute to enhance the attention of authorities for the role of science, technology, and innovation policies in Latin America.

José Antonio Ocampo
Member of the Board of Banco de la República
(Colombia's central bank) and professor
School of International and Public Affairs
Columbia University
New York, NY, USA

Contents

1 Influence of Science, Technology and Innovation on Urban Sustainability 1
William H. Alfonso Piña and Clara Inés Pardo Martínez
1.1 Introduction 1
1.2 Data and Method 7
1.3 Results 8
1.3.1 Comparative Analysis of the Main Success Factors and Lessons Learned in Different Regions from the Application of STI Guidelines 8
1.3.2 Characteristics of Trends in STI in the Selected Latin American Cities 11
1.4 Conclusions 16
References 17

2 Applying a New Methodology to Measure Investment in R&D and Science and Technology Activities: The Case of Colombia 21
Alexander Cotte Poveda, Jorge Andrade Parra, and Clara Inés Pardo Martínez
2.1 Introduction 21
2.2 Data and Methods 25
2.3 Results 27
2.3.1 Trends in STIA Expenditure in Colombia 27
2.3.2 Investments in STIA by Colombian Institutions 28

2.3.3 *Distribution of STIA Expenditure by Type of Activity* 28
2.3.4 *Governmental STIA Funding by Socioeconomic Objective* 30
2.4 *Conclusions* 32
References 33

3 Bioprospecting Model for a New Colombia Drug Discovery Initiative in the Pharmaceutical Industry 37
Juan Bueno and Sebastian Ritoré
3.1 *Introduction* 37
3.2 *The Biodiversity of Colombia* 39
3.3 *Sustainable Biocommerce: The Valuation of Life and Application of Biodiversity to Benefit the Prosperity of Colombia* 40
3.4 *Problem Tree Analysis: To Understand the Problem Is to Have the Solution* 43
3.5 *SWOT Analysis for Implementation of a Bioprospecting Program in Colombia* 45
3.5.1 *Market Study on the Biopharmaceutical Industry in Colombia* 45
3.6 *Risk Analysis and Risk Management in a Bioprospecting Program* 54
3.7 *Trends in Bioprospecting to Solve Problems in Innovation* 56
3.8 *Conclusions* 56
References 58

4 Patents for All: A Content Analysis of an Open-access Dataset of Colombian Patents 1930–2000 65
Julián David Cortés-Sánchez
4.1 *Introduction* 65
4.2 *Literature Review* 68
4.3 *Methods* 73
4.4 *Results* 74
4.5 *Discussion* 82
4.6 *Conclusions* 89
References 91

5 Open Access Policies in Science to Promote Science, Technology and Innovation in Emerging Economies. The Case of Colombia 95
Karen Isabel Cabrera Peña and Patricia Beatriz Márquez Rodríguez
5.1 Introduction 95
5.1.1 Literature Review 96
5.2 The Case of Colombia and Latin America 99
5.3 Methods 100
5.4 Results 101
5.4.1 Countries with Open Access Policies Associated with STI 101
5.4.2 Colombian Legal Framework on STI and Its Relation to Open Access 106
5.5 Conclusions 115
References 117

6 The Inclusion of Ecuadorian Women in Scientific-Technological Fields 121
Antonio A. Franco-Crespo, Fernando Herrera García, and Jessica Lin Gutiérrez
6.1 Introduction 121
6.2 Literature Review 123
6.3 Methods 131
6.4 Results 133
6.5 Conclusions 137
References 139

7 Effects of Expenditures in Science, Technology and R&D on Technical Change in Countries in Latin America and the Caribbean 143
Alexander Cotte Poveda and Clara Carolina Jimenez
7.1 Introduction 143
7.2 Data and Methodology 145
7.2.1 Data 145
7.2.2 Methodology 146
7.3 Results 147
7.4 Conclusions 150
References 154

8 Science, Technology and Water Access 157
Gustavo Correa Assmus
8.1 Introduction 157
8.2 Theoretical Framework 158
8.3 Data and Methods 160
8.4 Results 162
8.4.1 Business Characterization 162
8.4.2 Budget Participation 163
8.4.3 Inflationary Effect 163
8.4.4 Tax Participation 163
8.4.5 Training of Scientific Staff 164
8.4.6 Patent Development and Innovation 164
8.5 Discussion 165
8.6 Conclusions 166
References 167

9 The Ecuadorian Scientific Technological Policy Within the Framework of "Buen Vivir", a Dispute with the Orthodoxy of Development 169
Fernando Herrera García, Jessica Lin Gutiérrez, and Antonio A. Franco-Crespo
9.1 Introduction 169
9.2 Literature Review 170
9.3 Methodology 174
9.4 Results and Discussion 175
9.4.1 The Academy-Government Relations Regarding Higher Education Reform 178
9.4.2 Change of PS&T and Conceptions of the Relevant Participants 181
9.5 Conclusions 187
References 189

10 Against All Odds: Scientific Education, Scientific Competences and the Ideals of an Educated Country, Colombia 1995–2010 195
Gina G. Calderón and Edgar G. Eslava
10.1 Introduction 195

10.2 Literature Review 197
10.2.1 UNESCO on Scientific Education and Competences 198
10.2.2 European Union "Key Competences" 200
10.2.3 Competences in OECD DeSeCo Project 202
10.2.4 Competences and Scientific Competences in PISA 204
10.2.5 Studies About Scientific Competences in Colombian Scientific Education Policies 207
10.3 Methodology 208
10.4 Results 211
10.4.1 General Statistics 212
10.4.2 Clusters Analysis: Internal Coherence and Consistency 213
10.4.3 Survey: The Vision of the Colombian Teachers and Experts 216
10.5 Conclusions 218
References 219

11 Perception in Science, Technology and Innovation: A Comparative Review for the Years 2012 and 2015 from Colombian Surveys 225
Sylvia Patricia Fletscher Moreno and Clara Inés Pardo Martínez
11.1 Introduction 225
11.2 Research Method 227
11.3 Results and Implications 229
11.3.1 Socio-demographic Characteristics 230
11.3.2 Interest and Information S&T 230
11.3.3 Attitudes and Valuation 232
11.3.4 Social Appropriation of Science and Technology 233
11.3.5 Citizen Participation 234
11.3.6 Public Policies 235
11.3.7 Perception Indexes 236
11.4 Conclusions 242
References 244

12 Innovation, Entrepreneurship, and Technology-Based Clusters in Emerging Economies: How Can the Madrid Aerospace Cluster Reinforce the Mexican Aerospace Industry? 245
José Manuel Saiz-Alvarez and José Ramón Gutiérrez-Martín
12.1 Introduction 245
12.2 Literature Review 248
12.2.1 The Importance of Entrepreneurship-Based Innovation 248
12.2.2 The Development of Industrial Clusters 250
12.2.3 Technological Innovation and Endogenous Development 253
12.2.4 Innovation Processes in Sectoral Clusters 257
12.2.5 Cluster Location 258
12.2.6 The Querétaro Aerospace Valley (Mexico) 260
12.2.7 The INNOVA Model of Madrid Aerospace Cluster (Spain) 262
12.3 Method 263
12.3.1 Research Results 264
12.4 Conclusions 268
References 270

13 Strategies for Identification of Interdisciplinary Research Subjects: Approach, Tools, and Opportunities for Sustainable Development of Agricultural Economies 275
Wilmar Osorio Viana, Paola Andrea Calderón Cuartas, and Javier Mauricio Naranjo Vasco
13.1 Introduction 275
13.2 Literature Review 279
13.3 Methods 281
13.4 Results 286
13.5 Conclusions and Significance 293
References 293

14 An Analysis of Urban Public Policy Regarding Science, Technology and Innovation from the Perceptions of Stakeholders: A Case Study for a City of an Emerging Country 297
Alexander Cotte Poveda, Clara Carolina Jimenez, and Clara Inés Pardo Martínez
14.1 Introduction 297
14.2 Trends in Science, Technology and Innovation (STI) Policy in Bogotá 301
14.3 Methods 305
14.4 Results 309
14.5 Discussion 314
14.6 Conclusions 316
References 317

Glossary of Acronyms 319

Notes on Contributors

William H. Alfonso Piña Architect, with intensification in Architectural Design and Environmental Issues, Master in Theory of Architecture and holds a PhD. The professional performance experience has been oriented towards the management and research in land use, public space, heritage and coordination and development of urban and environmental studies, which includes the implementation of investment projects.

Jorge Andrade Parra Master in Economics, Master in Public Policy and Economist from the Universidad de los Andes. More than seven years of professional experience in the public sector. With advanced knowledge of public finances, public innovation in Colombia, public budgets, national competitiveness and economic information analysis, he has served as advisor to the Ministry of Finance.

Juan Bueno Researcher in Bioprospecting development and consulting. Chief Scientific Officer Fundación Centro de Investigación de Bioprospección y Biotecnología de la Biodiversidad Biolabb. Doctor in medicine, master's degree in biomedical sciences. Expert in design and implementation of anti-microbial platforms to evaluate natural products and synthesis drugs with the aim of optimizing in vitro antimicrobial activity.

Karen Isabel Cabrera Peña Assistant Professor of Law at Universidad del Norte, with a PhD in Law and expertise in computer and society, public law and legal process. Her research interests are public and legal administration, legal culture, civil and commercial matters. She has some publications related to open science, copyright, public policies, internet governance, among others.

Gina G. Calderón Licentiate in Biology and Education with an MSc in Science Didactics. Specialist in the development of integration, appropriation and evaluation projects of ICT in educational environments. Researcher and consultant in educative innovation, evaluation of educational models, contents design and Virtual Learning Environments. Works in the Education Innovation Office at Colombia's National Ministry of Education.

Paola Andrea Calderón Cuartas Environmental manager with an MSc in Agricultural Production Systems, Associate Professor in the Department of Environmental Engineering of Universidad Católica de Manizales (Colombia). Her research interests include management processes, culture and environmental education for sustainable production and consumption. She is leading a research project to understand and improve the environmental management systems used in Latin American universities.

Gustavo Correa Assmus PhD in Agrociencias, research professor at the Universidad de La Salle. His research interests are water, poverty and development in Colombia. He has experience and publications related to environment, education, environmental economic valuation, environmental services payment, sustainable management, Colombian natural parks, productivity and competitiveness, among others.

Julián David Cortés-Sánchez is resigned to perpetual learning. He is a distinguished professor at the Universidad del Rosario's School of Management (Colombia). He has been a teaching assistant at Universidad de Los Andes (Colombia), a junior researcher funded by Colciencias (Colombia), and a guest lecturer at the University of California, Berkeley.

Alexander Cotte Poveda PhD (c) in Economics from the Universität Göttingen in Germany, master's in Economics from the University of the Andes, Economist from the Central University. Research Associate of the Colombian Observatory of Science and Technology—OCyT. International Consultant on Science, Technology and Innovation for the World Bank. Research topics are research, development and innovation—R&D+I, among others.

Edgar G. Eslava BSc in Physics and Education, MSc and PhD in Philosophy. Teacher, researcher and consultant in the areas of international education and educational policies. Faculty member of the Philosophy and Letters department at Universidad Santo Tomas. Leads research projects in philosophy of science and technology, philosophy of education and integrated teaching.

Sylvia Patricia Fletscher Moreno Magister in Human Resources Management and Organizational Development, Universidad Externado de Colombia. Specialist in International Business and International Trade. Vast experience in project management, and in administrative, academic and financial management. Professor of Microeconomics, Economic Theory and Principles of Economics in Universidad Externado de Colombia.

Antonio A. Franco-Crespo Engineer in Electronics and Telecommunications, Specialist in Educational Management, Master in Business Management and holds a PhD in Administration (UASB). Professor at Escuela Politécnica Nacional in Ecuador. His lines of research are the impacts of science, technology and innovation in society; women, science and technology; and the market-environment-society relationship.

José Ramón Gutiérrez-Martín PhD in Business Sciences, Nebrija University, and executive MBA (EMBA), IESE Business School (University of Navarra). Full professor, consultant, and MBA, EMBA degree holder and Alumni Director, ESDEN Business School, Spain. He has also taught in Spain (Carlos III University) and Japan (IUJ, Niigata; and YCU, Yokohama) on entrepreneurship, finance, innovation, and business in Eastern Asia.

Fernando Herrera García Engineer in Electronics and Information Networks, Master in Development Economics and Doctor in Social Sciences from FLACSO Ecuador. Professor—Researcher at Escuela Politécnica Nacional in Ecuador. Member of the ESOCITE and of the "Ecuadorian Society of Social Studies of Science and Technology". Specialist in science, technology and innovation policy; and political economy of public policies.

Clara Carolina Jimenez Political scientist and Magister in Economics of Public Policies of the Universidad del Rosario. With skills in qualitative and quantitative analysis. With experience both in the Education Sector, and in NGO and Consulting. A researcher in scientific policy and innovation of the Colombian Observatory of Science and Technology.

Jessica Lin Gutiérrez Business Engineer, graduated at Escuela Politécnica Nacional in Ecuador. She is a product manager, and researches, selects, and promotes the development of products in the organization where she works, in Buenos Aires city. She is doing her master's degree in International Economic Relations at the University of Buenos Aires.

Patricia Beatriz Márquez Rodríguez Assistant Professor of Business School at Universidad del Norte. PhD in Integration of Information Technologies to organizations and Industrial Engineer from Universidad Politécnica de Valencia. Her research interests are innovation and entrepreneurships, innovation management, social innovation, value chain, among others, with publications, experience and consultancies in these topics.

Javier Mauricio Naranjo Vasco PhD in Chemical Engineering, Associate Professor in the Department of Environmental Engineering of Universidad Católica de Manizales (Colombia). An expert in the use of agro-food waste for biorefineries, he is now focusing on industrial, agro-industrial and urban solid waste valorization, taking into account issues on environmental education, urban environmental management and life cycle assessment.

Wilmar Osorio Viana PhD in Chemical Engineering, Associate Professor in the Department of Environmental Engineering of Universidad Católica de Manizales (Colombia). His main research interests are now focusing on the application of advanced technologies for the valorization of residues into high-tech products and the knowledge management innovation strategies required to increase social development in Latin America.

Clara Inés Pardo Martínez Food, Environmental and Sanitary Engineer, with a master's degree in Management, PhD in economics, and Postdoctorate in energy and climate studies. International leading auditor with experience in advising, consulting, research and teaching on issues related to environmental and sustainability. She is full professor at the Universidad del Rosario's School of Management in Bogotá, Colombia and Consultant of different national and international organizations.

Sebastian Ritoré is a photographer who works with the government of Quindio, Armenia, Colombia.

José Manuel Saiz-Alvarez PhD in Economics and Business Administration, Autonomous University of Madrid and a PhD in Sociology, the Pontifical University of Salamanca. GIEE faculty-researcher, EGADE Business School—Tecnologico de Monterrey (Mexico), and invited professor of different universities. Accredited by the Spain Agency for the Evaluation of Quality and Accreditation and by the Mexican System of Researchers (Mexico).

List of Figures

Fig. 1.1	The main indicators of Santiago de Chile related to economic performance and STI (patents are not indexed by number of patents per year)	12
Fig. 1.2	The main indicators of Mexico City related to economic performance and STI (patents are not indexed by number of patents per year)	14
Fig. 1.3	The main indicators of Bogotá related to economic performance and STI (patents are not indexed by number of patents per year)	15
Fig. 2.1	Trends in STIA as a percentage of GDP. Source: OCyT (2018). Note: *p* provisional data, *pr* preliminary data, *py* projected data	27
Fig. 2.2	Trends in STIA investment as a percentage of GDP. Source: OCyT (2018). Note: *p* provisional data, *pr* preliminary data, *py* projected data	30
Fig. 2.3	Investments in R&D by socioeconomic objective. Source: OCyT (2018). Note: *p* provisional data, *pr* preliminary data, *py* projected data	31
Fig. 2.4	Investments in STIA by socioeconomic objective. Source: OCyT (2018). Note: *p* provisional data, *pr* preliminary data, *py* projected data	32
Fig. 3.1	Problem tree analysis of bioprospecting and the biotechnology industry in Colombia	44
Fig. 3.2	Innovation trends in phytotherapeutics (number of products developed) (Superintendencia de Industria y Comercio-SIC, CIGEPI, 2015)	51

Fig. 3.3	Phytotherapeutic patent applicants in Colombia (Superintendencia de Industria y Comercio-SIC, CIGEPI, 2015)	51
Fig. 3.4	Types of patent applicants for phytotherapeutics in Colombia (SIC, CIGEPI, 2015)	52
Fig. 3.5	Companies applying for patents on phytotherapeutics in Colombia (Superintendencia de Industria y Comercio-SIC, CIGEPI, 2015)	52
Fig. 3.6	Trademarks registered in Colombia by leading patent applicants on phytotherapeutic (Superintendencia de Industria y Comercio-SIC, CIGEPI, 2015)	53
Fig. 4.1	Dependency rate, 2006–2015. Source: OCyT, 2015, p. 120	70
Fig. 4.2	Self-sufficiency rate, 2006–2015. Source: OCyT, 2015, p. 121	71
Fig. 4.3	Invention coefficient, 2006–2015. Source: Author, based on OCyT, 2015, p. 122	71
Fig. 4.4	Colombians' patent application and granted patents rates in international offices, 2006–2015. Source: Author, based on OCyT, 2015, p. 123	72
Fig. 4.5	Number of patents granted in Colombia in 1930–2000. Source: The author based on Mayor, 2005	74
Fig. 4.6	Invention coefficient 1985–1998. Source: The author based on Mayor, 2005, and the National Statistics Department (DANE), 1985–1998	75
Fig. 4.7	Publication of patents by type of inventor. Source: The author based on Mayor, 2005	75
Fig. 4.8	Content analysis of word frequency in patent titles. Source: The author based on Mayor (2005) and analyzed using Voyant Tools	76
Fig. 4.9	Ratio of the terms in patent names in 1930–1953, 1954–1977, and 1978–2000. Source: The author based on Mayor (2005) and analyzed using Voyant Tools	76
Fig. 4.10	Collocation graph of keywords in patent names in 1930–1953. Source: The author based on Mayor (2005), and analyzed using Voyant Tools	77
Fig. 4.11	Collocation graph of keywords in patent names in 1954–1977. Source: The author based on Mayor (2005), and analyzed using Voyant Tools	78
Fig. 4.12	Collocation graph of keywords in patent names in 1978–2000. Source: The author based on Mayor (2005), and analyzed using Voyant Tools	78

Fig. 4.13 Collocation graph of keywords in patent names granted in cities in the Andean (left) and Caribbean (right) regions. Source: The author based on Mayor (2005), and analyzed using Voyant Tools 79

Fig. 4.14 Collocation graph of keywords in patent titles granted to companies. Source: The author based on Mayor (2005), and analyzed using Voyant Tools 80

Fig. 4.15 Collocation graph of keywords in patent titles granted to groups of people. Source: The author based on Mayor (2005), and analyzed using Voyant Tools 80

Fig. 4.16 Collocation graph of keywords in patent titles granted to individual male inventors. Source: The author based on Mayor (2005), and analyzed using Voyant Tools 81

Fig. 4.17 Collocation graph of keywords in patent titles granted to individual female inventors. Source: The author based on Mayor (2005), and analyzed using Voyant Tools 81

Fig. 4.18 Number of patents by IPC section. Source: The author based on Mayor (2005) and WIPO (2017a) 83

Fig. 4.19 Top five IPC sections from 1930 to 2000 and patent application by field of technology 2017. Source: The author based on Mayor (2005) and WIPO (2017a) 88

Fig. 5.1 Evolution of GDP (US$ millions) vs. R&D Investment (as a GDP Percentage) in Colombia, 1990–2015 113

Fig. 5.2 Evolution of GDP (US$ millions) vs. R&D investment (as a GDP percentage) in Colombia, 1990–2015 114

Fig. 5.3 Number of Open Access publications in Scopus. Colombia vs. Worldwide 1996–2016 114

Fig. 7.1 Technical change, research and development, and science and technology in Latin America and the Caribbean 151

Fig. 7.2 Technical change, science publications and patents in Latin America and the Caribbean 152

Fig. 7.3 Technical change, education and expenditure in technology and communications in Latin America and the Caribbean 153

Fig. 9.1 Higher education system and science and technology system before 2010. CONESUP (Consejo Nacional de Educación Superior), CONEA (Consejo Nacional de Evaluación y Acreditación de la Educación Superior) 181

Fig. 9.2 Higher education system and science and technology system after 2010. CES (Consejo de Educación Superior), CEACCES (Consejo de Evaluación, Acreditación y Aseguramiento de la Calidad de la Educación Superior) 182

Fig. 11.1	Do you consider that science or technology is made in Colombia? Source: SPPS&T, 2012 and 2015 OCyT	229
Fig. 11.2	Media by which the respondent is informed about science and technology. Source: SPPS&T, 2012 and 2015 OCyT	231
Fig. 11.3	Problems that could be solved with innovation. Source: SPPS&T, 2012 and 2015 OCyT	233
Fig. 11.4	Activities related to science and technology done by respondents. Source: SPPS&T, 2012 and 2015 OCyT	234
Fig. 11.5	Solutions generated by science and technology according to respondents in both surveys. Source: SPPS&T, 2012 and 2015 OCyT	236
Fig. 12.1	Growth axis in the aerospace industry. Source: Authors	247
Fig. 13.1	Colombian model for science, technology, and innovation (Adapted from CONPES, DNP, 2015)	279
Fig. 13.2	RG-TED research lines and its year of creation	282
Fig. 13.3	Particular research topics emerged from the RG-TED research lines	283
Fig. 13.4	RG-TED historic research subjects ordered by their global projection indicator	288
Fig. 13.5	Conceptual map for the emergence of interdisciplinarity in RG-TED	289
Fig. 13.6	Conceptual frame in order to understand the group behavior and performance dynamics	291
Fig. 13.7	A general strategy for the knowledge management for research groups in LA&C countries	292
Fig. 14.1	Potential indirect influence and dependence map	310
Fig. 14.2	Relational analysis for role the State, universities and enterprise	313
Fig. 14.3	Value chain in the formulation of STI policy	315

List of Pictures

Picture 1	The *páramo* ecosystem represents 17% of Colombia's floral diversity, yet covers only 2.5% of the continental territory (Marín & Parra, 2015)	40
Picture 2	Any biotrade initiative must be registered under the Nagoya protocol, so that the benefits are shared equitably in the communities affected (Kang, Jung, & Ryu, 2015)	41

List of Tables

Table 1.1	Comparative analysis related to success factors and lessons learned from STI guidelines	9
Table 2.1	Main studies of the relationship between R&D and economic growth and development	23
Table 2.2	The main trends in STIA investments by institutions, 2009–2017	29
Table 3.1	SWOT analysis for bioprospecting and biotechnology enterprises in Colombia	46
Table 3.2	Drugs under development by the pharmaceutical industry, 2011	48
Table 3.3	Top 10 pharmaceutical companies by sales (billions of dollars)	48
Table 3.4	Concentration of drug market in Colombia, 1993–2007 (Vásquez Velásquez et al., 2010)	54
Table 5.1	Countries with Open Access policies associated with STI	102
Table 5.2	Policy strategies to implement in countries with emerging economies (Colombia)	116
Table 6.1	The data of the number of publications made by Ecuador and its neighboring countries (2007–2016)	128
Table 6.2	The data of the number of publications made by Ecuador and its neighboring countries (2007–2016)	128
Table 6.3	The percentage of women's participation in the academic field	129
Table 6.4	The number of women who have traveled abroad for studies	130
Table 6.5	Recurring codes	133
Table 6.6	Categories	134

Table 7.1	Dependent GDP	146
Table 7.2	Fixed-effects estimations of technical change	148
Table 7.3	Fixed-effects with Driscoll and Kraay standard errors estimations of technical change	149
Table 11.1	Structure of survey on public perception of S&T	228
Table 11.2	Comparative results of the 12-perception index calculated from the perception surveys of the years 2012 and 2015 at the national level	237
Table 12.1	Firms belonging to the Querétaro Aerospace Valley (excluded SMEs)	260
Table 12.2	Descriptive statistics of selected variables from the sample	264
Table 12.3	Matrix of correlations	265
Table 12.4	KMO and Bartlett tests	266
Table 12.5	Commonalities	266
Table 13.1	Some countries in the Cornell University et al. 2017 ranking (adapted)	277
Table 13.2	Level of interdisciplinary treatment of a research topic and interdisciplinarity indicator	284
Table 13.3	Contribution scale of a research topic to the area of knowledge and complexity indicator	285
Table 13.4	Prioritized sustainable development goals for RG-TED/ context and congruence indicator	285
Table 13.5	RG-TED historic research subjects and indicators for knowledge management	287
Table 13.6	Key factors of the interaction in external couplings for RG-TED	292
Table 14.1	The current scheme of the STI policy in Bogotá	303
Table 14.2	Interviews and focus groups carried out within the framework of the project to determine the main factors to be considered in an STI policy	306
Table 14.3	List of factors obtained from interviews and focus groups	308
Table 14.4	Vester Matrix	309
Table 14.5	Influence matrix factor typology	309

CHAPTER 1

Influence of Science, Technology and Innovation on Urban Sustainability

William H. Alfonso Piña and Clara Inés Pardo Martínez

1.1 Introduction

Science, technology and innovation (STI) are fundamental to promoting sustainable development in cities, taking into account the fact that more than 90% growth in urban populations is expected in the next 30 years; it is important to determine what are the best conditions to maintain adequate infrastructure, reduce the use of resources, and prevent environmental deterioration and the risk of natural disasters, and STI have an important role because they can provide different solutions to improve the quality of life in urban areas (UNCTAD, 2013).

Cities that are global knowledge leaders are normally characterized by a high concentration of STI resources that attract the best human resources, effective researchers and institutions that generate quality research and

W. H. Alfonso Piña (✉)
Universidad del Rosario, Bogotá, Colombia
e-mail: william.alfonso@urosario.edu.co

C. I. Pardo Martínez
School of Management, Universidad del Rosario, Bogotá, Colombia
e-mail: clara.pardo@urosario.edu.co

C. I. Pardo Martínez et al. (eds.), *Analysis of Science, Technology, and Innovation in Emerging Economies*,
https://doi.org/10.1007/978-3-030-13578-2_1

social impact, a friendly commercial environment with an open culture, a knowledge-based economy conducive to the growth of emerging industries and close links with global STI networks generating growth development through international collaborations (Advertisement Feature, 2017).

Factors that promote growth and development in the regions around the world have close relationships with investments and the strengthening of research and development (R&D) in different socioeconomic contexts and local wealth, indicating the regional value added by a complex and holistic STI approach, which allows for a decrease in inequality and a strengthening of land-use planning based on knowledge that is key to smart cities (Dettori, Marrocu, & Paci, 2012). This approach has been applied by different regions generating transformations in the flows of investments, information and services in the main commerce hubs of (by order of importance) London, New York and Tokyo, followed by Madrid, Milan and Shanghai, as second hub; Latin American cities have begun to work to apply STI in urban management with the aim of connecting to the flows of global commerce (Caicedo Aspril, 2011).

Considering that most urban regions share a large number of similarities and productive dynamics, a strategic approach is needed to help differentiate between regions, beyond landscape and cultural aspects, through increased investment in R&D, which improves the value added and productivity. In Latin America, this strategy for global competitiveness is known as investment in STI.

In the cities, technology can improve urban planning and governance by using different strategies to improve transportation and water and energy supplies, localize settlements with lower risks and better localize urban infrastructure.

This study starts from the conception of a correlation among the concepts of R&D, regional differentiation and development. The existence of such a correlation is clear from the fact that the current global dynamics have shown that what generates more added value to cities and regions is their investment in STI, which helps to overcome the homogenization of productivity between urban regions.

The main argument in favour of STI as a key to differentiated regional development is based on the fact that regional agglomeration provides the best context for a knowledge economy (Hudson, 1999) and for the creation and dissemination of knowledge as well as for learning (Valdez-Lafarga & León-Balderrama, 2015). Moreover, innovation plays a predominant role in the postulates of Porter (2000) to sustain competitiveness. The author argues that, for the specific case of a region, prosperity is linked to the region's competitive advantage, and the first driving force of competitive advantage is innovation that depends on the results and process of knowledge generation through science and technology.

Camagni (2003, p. 8) states that, from an urban economy perspective, the explanatory arguments of the competitive advantages of large cities and urban agglomerations always attach special importance to the effects derived from innovation, which is a dimension that generates economies of scale derived from the indivisibility of some processes and is the fundamental cause of the emergence of positive externalities. Therefore, a large number of metropolitan regions of the world are those that, with their stockpiling of physical, human capital, finance and knowledge, generate the largest volume of localization and urbanization economies and constitute an attractor for all types of activities, especially those with high STI content as a key element in the sophistication of markets and higher growth.

The analysis of STI as a competitiveness factor highlights the relationship among region, knowledge, innovation and diffusion, as well as the spatial dimension of the technical change supported by the construction of local and/or national innovation systems (Ramirez J. & Parra-Peña S., 2010), and demonstrates that the notions of technical change, innovation and competitiveness are closely related to development (Cimoli, Carlos, & Annalisa, 2005).

Globalization has opened up a wide range of opportunities for the productive sectors in developing countries to be competitive and exploit forms of improved growth and progress. However, a basic requirement for development is that the different production sectors increase their technological standards (Chadha, 2007, p. 415) to a great extent; it is now essential for industries to expand their STI efforts in a liberalized and globalized environment to be able to respond successfully to the current competition and production dynamics.

All of the above has generated increasing recognition among regional authorities that economic growth and the competitiveness of their regions depend to a large extent on the ability of the companies in these regions to innovate, which is why offering appropriate support to help local companies to be more innovative and competitive is now a key factor in regional policy (Cooke & Memedovic, 2003, p. 8).

Second, STI is not only crucial for regional economic competitiveness but also for the growth of basic productivity. The findings suggest that expenditures on R&D, industrial diversity and the distribution of human capital endowments can have an important effect on the generation of innovation and productivity (Mukim, 2012, p. 357).

Gordon and McCann (2005, p. 4) offer a vision of the relationship between innovation and regional development in which they focus on the role of agglomeration economies in the promotion of localized learning processes, such as indirect effects or other transfers of information, through which it is possible to identify the amount of benefits to localized regional companies as a result of the development of new products, services and processes.

Likewise, there is an extensive literature regarding the analysis of technical change and endogenous economic growth, where authors indicate that innovation plays a transcendental role because it becomes a catalyst for growth through the progressive increase and effectiveness of the factors of production. For example, Coe, Helpman, and Hoffmaister (1997) empirically demonstrate the positive impact of R&D on productivity, and Solow (1956) emphasizes the technological differences that explain a large part of countries' productivity differentials.

In this way, regional productivity is also positively affected by the activity of STI and the improvement of public infrastructure in neighbouring regions given that technological knowledge, which is created and accumulated through STI activities, improves the production and diffusion of innovations and subsequently promotes the growth of productivity and competitiveness (Bronzini & Piselli, 2009).

At the global level, it is assumed that increases in R&D investment in any territory will lead to greater development, knowledge generation and innovation, which in turn will lead to greater economic growth. In Latin America, it is also assumed that the focus on and investment in STI can be one of the main drivers of economic growth, both nationally and regionally. Indeed, Goschin (2014, p. 24) has argued that any region that invests sufficient resources and is effectively involved in STI has the potential to achieve economic growth and that, in fact, there is a bidirectional connection between this focus and economic growth. A key place for regional development (measured by regional GDP per capita) often includes determinants of regional investments in R&D and the knowledge flows of research institutes and universities that are usually located together geographically and stimulate productivity and regional growth.

In addition, studies that consider R&D in combination with local wealth and socioeconomic conditions indicate that regional economic growth and innovation are also strongly mediated by how local social and economic conditions and, in particular, strong human capital growth configure the returns of investment in R&D (Usai, 2011).

In spite of the above, this hypothesis has already been questioned by the literature in various territories (Rodríguez-Pose & Villareal Peralta, 2015), and there are studies with results that suggest that, although the expenses of STI have an impact on the growth of innovation, the grouping of innovative activity and input factors associated with the advantages of a location are also important (Mukim, 2012, p. 359) and the promotion of STI activity can be considered a weak instrument even to reduce regional disparities (Bronzini & Piselli, 2009, p. 197). Some authors have expressed doubts about the infallibility of the correlation between R&D, growth and development, stressing that the ability to appropriate this knowledge should also be introduced into the equation so that knowledge can be converted into growth. These authors also suggest that the correlation depends on many other factors that allow for the strengthening of the socioeconomic environment so that it is conducive to innovation and entrepreneurship (Goschin, 2014, p. 25).

Fourth, and closely related to the variable of economic growth, questions arise regarding the role of STI in reducing regional inequalities. Although some research has found that the various R&D expenses do not affect the level of disparity in economic growth (Aritenang, 2013, p. 155), for several decades authors such as Posner (1961) and Krugman (1979) have identified that, in the analyses between regions, the presence or lack of innovation can affect the rates of differential growth in particular, and imitative or innovative management can explain different levels of development between countries or regions, such as the "technological gap" or even the "north-south" asymmetry (Noronha Vaz, Noronha Vaz, & Nijkamp, 2013, p. 8).

More recent studies have also shown that rich regions are able to take advantage of their R&D activity, while poorer regions would benefit more from imitation. This postulate is based on the propensity for a high intensity of specific R&D in the regions that have overcome a certain level of development (Braconier, 2000). In fact, Goschin (2014) notes that in the R&D-based growth models the propensity to invest in STI is related to the size of the economy since the larger regions, which are characterized by larger markets and more resources, are more motivated to carry out R&D activities, which in turn promote more rapid processes of innovation and economic growth.

At the level of scales, Méndez, García Palomares, and Michelini (2007, p. 5) identify the importance of the innovation process in intermediate cities to generate, disseminate and exchange knowledge and take advantage of patrimonial resources, which allow the consolidation of innovative environments which are capable of promoting better development as well as greater growth. In this sense, it is convenient to take into account the fact that the different capacities that the regions may present do not necessarily imply that the most backward regions do not generate innovation processes, but that they can develop different innovation dynamics with less radical processes and less specialization or more processes that are focused on aspects related to business demography (Aguila, Garrido Yserte, & Gallo Rivera, 2010, p. 28). Decision makers can generate great comparative advantages and benefits from having better knowledge of those critical factors that maintain regional disparities in STI, as well as the success factors and the mechanisms through which the development of research and innovation can contribute to the differentiated economic growth of regions, whereas those with less capacity for innovation do not necessarily have the lowest economic indicators.

In the current phase of globalization, where regional specialization alone would strengthen the comparative advantages of a territory in world markets, it is necessary that the industrial development of regions be conceived on a solid basis. In addition, it is essential that regions not only specialize in certain types of industries but also increase their technological standards to improve their comparative and competitive advantage. For many, this can be achieved by allocating a larger proportion of the state's budgetary spending to public—private partnerships and the production sector in R&D processes (Chadha, 2007, p. 424).

Specialization plays an important role in regional development, which is evidenced by, among other things, the fact that higher incomes in rich regions constitute an incentive for buyers to search for high-quality and differentiated products that are the result of intense R&D activities, which leads to a high correlation between per capita R&D expenditures and per capita GDP (Goschin, 2014, p. 25), as well as the dependence of innovation on support advantages, such as the presence of industry networks, the concentration of R&D activity and the presence of a large set of service activities with multiple specializations (Mukim, 2012, p. 358).

Since the mid-1980s the importance of innovation processes for modern economic development has been highlighted (Valdez-Lafarga & León-Balderrama, 2015, p. 518). From this it is possible to conclude that, beyond the popularity of the STI approach to providing a narrative about the intangible dimensions of local economic development and the processes of the circulation of knowledge and learning at an apparently more manageable regional scale, it is clear that different empirical results suggest that regional investments in R&D are strong predictors of local innovative performance, and their economic and social effects are much broader in the long term. However, these activities are inadequately connected with regional human capital resources (Crescenzi & Jaax, 2017, p. 1).

Given this background, this study seeks to advance the analysis of the relationship between economic variables and STI performance in order to determine the critical factors and the success factors of the socioeconomic environment of each city that has allowed STI to be a driver of economic growth and improvement for the quality of life. STI can lead to greater regional growth. Therefore, increases in the financing of these activities, as well as their balanced territorial distribution, should be used as vectors to improve the economic performance of the regions, increase the competitiveness of the regional economies and reduce existing disparities. The remainder of this chapter is organized as follows. Section 1.2 provides a description of the data and the method. Section 1.3 describes the main results and discussion. Section 1.4 offers some concluding remarks.

1.2 Data and Method

This study begins by providing background on the relationship and importance of STI in different regions, such as London, the Baltic region and Mexico, through a literature review identifying the main success factors and lessons learned for a comparative analysis.

Next, the study selects different cities in Latin America according to available data to determine the relationship between STI performance and economic variables. Santiago de Chile, Mexico City and Bogotá, which are considered the three smartest cities in the region, were selected for this study (Cohen, 2014a).

For each city selected, certain attributes are analysed, such as GDP (measured as millions of US$ 2010), labour productivity (which is the ratio between GDP and total employment), GDP per capita (US$ 2010) and total patents (number). The main data sources used in this study to calculate the variables for the three cities come from each city's respective statistics offices, the Economic Commission for Latin America and the Caribbean (ECLAC) database, OECD statistics and the Colombian Observatory of Science and Technology.

1.3 Results

This section provides the analysis of the main success factors and the lessons learned in the selected cities based on the STI initiatives, and describes the main economic and STI indicators in the Latin American cities selected for this study with the aim of describing the relationship among the economic variables, STI and smart cities processes, taking into account the available data as explained in the previous section.

1.3.1 Comparative Analysis of the Main Success Factors and Lessons Learned in Different Regions from the Application of STI Guidelines

Recently, the factors that promote socioeconomic development have been reassessed, and information, communication and innovation have been found to be central factors in a knowledge society. In this context, regions and cities generate knowledge and innovation as key elements in the creation, absorption and diffusion processes that allow the consolidation of STI as a main factor in production and development (Ligenzowska, 2015).

Cities are complex, market-driven systems with economic, social and technological relationships; in cities, institutional structures are different and depend on size, specialization and their role in the region. In this context, cities with greater opportunities have adaptive capacity and take advantage of technology, economic and market opportunities that are important factors in economic development. Furthermore, as new sectors, technologies and institutions appear, they replace old processes and generate a dynamic urban competitiveness through novelty and innovation (Office of the Deputy Prime Minister (ODPM, 2006). Table 1.1 shows a comparative analysis of the success factors and lessons learned from the STI guidelines of three cities.

These cases demonstrate that innovation is the basis to build competitive advantage around the world, generating regional development through new employment opportunities and strengthening markets at the regional level to attract new investors, cooperation opportunities with other regions and entities, and knowledge transfers, and to use experiences to generate new ideas and more competitive human capital (Ligenzowska, 2015).

Table 1.1 Comparative analysis related to success factors and lessons learned from STI guidelines

City or Region	*Success factors*	*Lessons learned*
London	• Support for innovative businesses taking into account the scale and scope of labour markets and the demand of customers, companies and the public sector.	• It is important to recognize the historical trajectory of every city.
	• To promote knowledge intensive business services (KIBS) associated with the academy and that are concentrated in this city.	• Formulation of STI policies should adapt to the environment and context of the city.
	• Economic transformation to generate conditions for the innovation of financial and business services to attract KIBS and develop national and international service centres.	• Effective policies should include bottom-up and top-down strategies.
	• To understand how a city maintains its competitive advantage and how it responds to technological change, demand, competences, policy regimes and other issues (Wood, 2009).	• Application of policies requires power and resources.
		• Need to improve strategies to finance and increase resources that guarantee adequate application of STI in all structures of cities.
Baltic region	• Quality in public management and capacity of political leadership to plan and implement innovative projects on a large scale.	• Need to strengthen the coordination between national and regional policies.
	• Innovation systems based on the third and fourth helix, generating stable economies, and universities that develop R&D activities are a resource for high technology industries.	• Need to separate the innovation strategies for the public and private sectors, taking into account the fact that their objectives are different.
	• Communication and dependencies between actors related to STI, generating autonomy in the managing entities.	• Effective communication among entities is needed to enhance capacities for innovation at the broadest level in the countries according to the requirements.

(*continued*)

Table 1.1 (continued)

City or Region	*Success factors*	*Lessons learned*
	• Implementation of a socioeconomic model based on increasing rates of innovation as a long-term strategy to stimulate innovation at the national and regional levels.	• Need to strengthen human resources related to STI activities and processes in all regions.
	• Investments in R&D are over 3% of GDP, and two thirds of those investments correspond to the private production sector.	• Knowledge exchange is generated in urban areas, and it is important to determine strategies for peripheral and remote regions. • Continuity is needed for technological policies, industrial clusters, technology incubation, technical advice and the development of research networks, among others (Goschin, 2014, Crescenzi & Rodríguez-Pose, 2013).
Mexico	• The openness of the economy to commerce generated higher competencies and the need to improve the productivity and innovation capacities of companies and businesses in a global context.	• The percentage of public investment in R&D is concentrated on basic research, but the preference should be applied research.
	• An increase in resources to promote R&D at all levels.	• Public and private institutions that participate in the generation of STI continue to operate independently without collaborating to maximize yields from R&D investments, requiring greater efforts and more work.
	• A national system of innovation approach to public policies that considers multiple network agents that integrate different public and private stakeholders to generate, assimilate, disseminate and apply new technologies and knowledge.	• Although the decentralization of the STI Mexican system has advanced rapidly, the majority of activities are concentrated in a single organization, which generates a bottleneck with regard to management.
	• The generation of indirect effects of knowledge through the dissemination of knowledge from where it is generated to neighbouring geographical areas (Rodríguez-Pose & Villareal Peralta, 2015).	• A geographic approach generates differences with regard to innovation capacities. Fifty per cent of the total R&D expenditures in the country are concentrated in eight Mexican states (Capdevielle & Dutrénit, 2012).

Moreover, STI must be analysed in cities from a holistic perspective, and a territorial analysis of STI policies should further the establishment of adequate policies in emerging countries according to their contexts.

Attractive cities for investments are located in countries with solid economies, efficient infrastructure and high levels of STI; it is essential to understand the actual situation of cities or regions with regard to their priorities, potential projects, resources, strategies and STI system features, among others, so that decision makers can receive and consolidate reliable information to analyse the comparative advantages and development prospects of every city or region (Collazos Rodriguez & Londoño Martinez, 2014).

The STI approach determines the future of an economy with regard to its commercial applications by generating high-quality knowledge and innovation through universities, the dissemination of scientific papers and the international dimensions of patent applications (OMPI, 2015), which demonstrates that STI is vital for prosperity and security.

Governments play an important role in stimulating STI that generates progress in the economy, social welfare, culture, health and other sectors. Different governmental and non-governmental institutions and public and private organizations should work together to resolve key problems and develop new products and services at a commercial scale to achieve better returns on STI investments.

1.3.2 Characteristics of Trends in STI in the Selected Latin American Cities

In the Latin American environment, STI investments are very low and there is too much lag between some regions and others, although it has been understood that the STI path is a strategy that allows for the promotion of the differentiated development of regions and therefore competitiveness. The incorporation of STI into urban management policies is recent and even investment resources are limited. This section shows the analysis of the Latin American cities selected for this study with the aim of describing STI trends according to certain indicators, taking into account the available data as explained in the previous section.

Santiago de Chile. This city has more than six million inhabitants or approximately 40% of Chile's national population and is considered the best city for doing business. The city has the lowest corruption rate in the region, the price of road tolls is automatized according to traffic, the city has 140% mobile phone penetration, and there was a 432% increase in mobile internet connections between 2009 and 2015 with different initiatives such as Smart Parking Entel and Chilectra that work seven programmes: electrical mobility technologies, domotics operation locations, photovoltaic generation from solar energy, remote measurement, electrical network automation, public LED lighting and telesupervision. The city also encourages entrepreneurship through programmes such as Start-Up Chile and programmes from institutions, universities and state agencies to generate a smarter city and execute different pilot projects across the city (Alfonso, 2014; Cartajena, 2016).

The roadmap for 2015–2025 includes the following topics (Cartajena, 2016): the specialization of the technology sector (start-ups), the specialization of human capital (data scientists and the use of information and communication technologies (ICT)), the interoperability of information (open data), the quality of digital infrastructure (smart industries and industrial internet) and the digitalisation of industries (big data, data mining and apps for smart cities). These elements are key to guaranteeing the development of a smart city in Santiago de Chile.

Figure 1.1 shows the main indicators of Santiago de Chile related to economic performance and STI; these indicate that there is a direct relationship between economic growth and new creations (measured by total number of patents). On average, this city generates approximately 65% of all patents in the country and contributes 46% of the country's GDP.

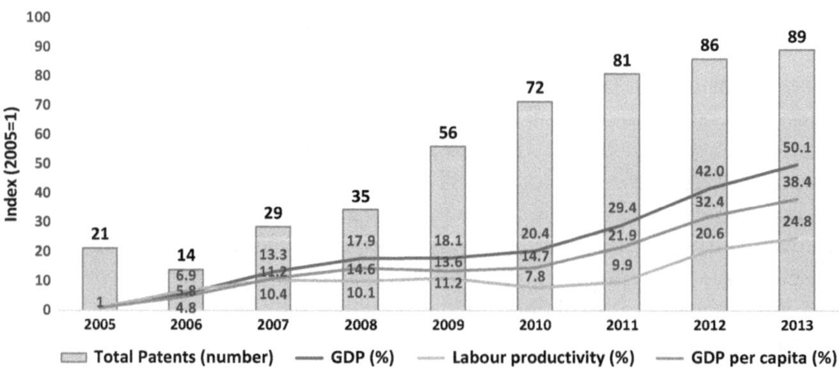

Fig. 1.1 The main indicators of Santiago de Chile related to economic performance and STI (patents are not indexed by number of patents per year)

However, STI investments are low at less than 0.40%. It is important to establish strategies to increase these investments with the aim of completing smart city projects and increasing integration with and investments by different stakeholders, particularly the production sector, universities and government, taking into account the requirements of urban areas.

Mexico City. The population of the metropolitan area is more than 20 million inhabitants. The city is characterized as a pioneer and generates incentives for the construction of intelligent and sustainable buildings, some of which absorb pollution, through the investigation of new green and cleaner technologies, digital research and open data initiatives by universities. The city has approximately 4000 public bikes for lease, soon expected to grow to 6000, which has led to the reduction of 232 tonnes of carbon dioxide in its first three years of operation (Alfonso, 2014; Cartajena, 2016).

The urban population of this city contributes to the improvement of urban conditions through small local projects characterized by a conscious urban plan that generates a more sustainable and smart management plan for the whole region with diverse programmes, such as a climate action plan to achieve a low-carbon society by reducing carbon emissions, the creation of a resilience department to prevent and control floods and earthquakes through disaster and environmental monitoring, and a traffic management and incident department with 20,000 sensors and cameras around the city complementing a smart grid system to reduce electricity theft (Mexico City Government, 2016).

The main STI objective of this city is to generate solutions to urban problems using technology facilitating the lives and productive activities of citizens through new tools based on collective intelligence and social and collaborative processes (SEDECO, 2016).

Figure 1.2 describes the main recent economic STI indicators for Mexico City. These indicators show growth, especially for indicators related to GDP and patents, which have maintained an increasing trend. On average, this city invests up to more than 0.50% of its GDP in R&D and generates approximately 25% of the patents in the country. However, this city has the highest concentration of STI capacities, which allows for the consolidation of a smart city and the achievement of better results when solving urban problems, taking into account the particularities of geography and environment. Public policies should continue promoting STI investments to develop new processes to generate new knowledge, technology and innovation for economic and social progress with instru-

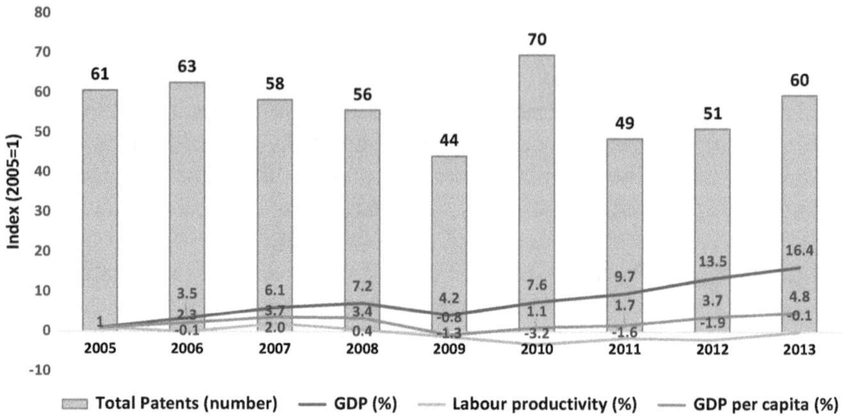

Fig. 1.2 The main indicators of Mexico City related to economic performance and STI (patents are not indexed by number of patents per year)

ments to motivate greater participation in projects by different stakeholders with public and private resources (CONACYT, 2016).

Bogotá. This city has nearly 10 million inhabitants and seeks to improve its congestion problems through the use of fast transit buses, which are used by 1.65 million people per day. The bus corridors have interconnected bicycle paths and have begun to integrate electric taxis in the partnership.

Bogotá has concentrated its efforts on improving mobility, applying smart transportation systems that allow for real-time monitoring and decision making with synchronized traffic lights and electronic toll collection, access to and dissemination of ICT through connectivity, and digital development and effective appropriation of ICT to improve education, entrepreneurships and open data to guarantee transparency and efficiency in public management (Bogotá Government, 2016).

Bogotá's smart city initiative seeks to improve its competitiveness index, increase international investments (tourism, companies and qualified personnel), attract new markets, promote local companies, generate knowledge for productive processes and improve the welfare of the population with the aim of guaranteeing sustainable development.

Figure 1.3 indicates the main STI indicators of Bogotá related to economic performance, demonstrating a direct relationship between economic growth and STI performance. This city is where the country is concentrating its STI investments and new knowledge. It is important to analyse priorities to strengthen the resolution of problems, taking into

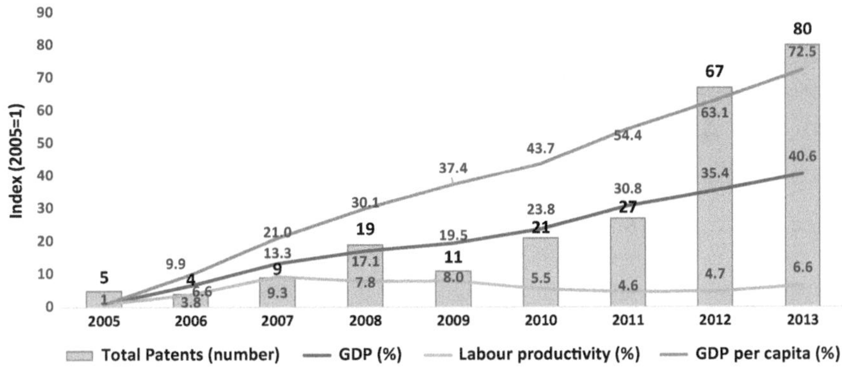

Fig. 1.3 The main indicators of Bogotá related to economic performance and STI (patents are not indexed by number of patents per year)

account STI and how regional inequalities can be decreased to promote integral development in the entire country.

STI have a direct relationship with economic factors; it is important to strengthen policies and programmes that improve the application of STI results with the aim of finding adequate solutions to improve sustainable development and respond to the increasing population trend in cities.

Strategies to improve STI should include a long-term formulation, the facilitation of the implementation of diverse programmes including regulations, the alignment of programmes and instruments with urban planning, the construction of an STI culture, the promotion of education and the sustainable development of urban areas.

It is important to develop adequate indicators to measure STI trends, taking into account the following features (Cohen, 2014b): (1) Government (online services, infrastructure and open government), (2) Citizens (culture and well-being, safety, health and technological infrastructure), (3) Mobility (efficient transport, multimodal access and technological infrastructure), (4) Economy (entrepreneurship and innovation, productivity, and local and global connections), (5) Society (inclusion, education and creativity) and (6) Environment (smart buildings, source management and sustainable urban planning). An adequate STI measurement system is key to achieving a knowledge society in the Latin American context with the aim of becoming smart cities based on STI, which increase productivity, economic growth, sustainable development and welfare.

1.4 Conclusions

This chapter analysed the relationship between economic performance and STI in three Latin American cities (Santiago de Chile, Mexico City and Bogotá), determining the direct relationship between these two factors. The STI trend has been maintained over recent decades in these cities, which has not allowed for adequate knowledge generation and effective innovation processes in some contexts. The comparison of the cities analysed shows that it is important to develop STI policies that promote solutions to the main problems of the region.

Results indicate that cities with better economic performance have achieved a higher number of patents as in the case of Santiago de Chile and Bogotá, indicating that higher economic growth is an important factor to increase STI results in cities of emerging economies, where labour productivity also has a direct relationship with STI results, which is important in the promotion of adequate economic and STI policies in urban contexts.

The identification of the trends in these cities, regarding the use of a regional differentiation approach to investments in STI, within the framework of the internationalization of the Latin American city helps reduce inequality gaps and improve productivity and management and sustainable development in a knowledge society, taking into account the fact that smart cities are a current topic that can provide several benefits to the inhabitants of Latin American cities.

STI are the basis for building competitive advantages around the world, generating regional development through new employment opportunities and strengthening markets at the regional level to attract new investors, cooperation opportunities with other regions and entities, and knowledge transfers, and to use these experiences to generate new ideas and more competitive human capital. This requires a holistic analysis of urban regions and STI policies. Such analyses should continue in emerging economies to establish adequate policies according to the contexts.

In Latin American cities, according to ECLAC (2016), it is important to increase investments in R&D to generate structural changes, which in turn encourage more STI, to promote the relationship among government, universities and the production sector to generate new knowledge and solutions to problems, to produce more knowledge and technology-intensive activities and sectors in the industrial structure together with greater knowledge spill-overs and production chains that enable the cities to grow their economies and improve their populations' living standards, and to encourage digital technologies that are key factors in productivity and competitiveness, among other things. STI is a virtuous cycle of growth that reflects increasing knowledge-based added value, development and social equality.

References

Advertisement Feature. (2017). *Measuring the Growth of Science and Technology Innovation in Shanghai*. Retrieved from https://www.natureindex.com/pdf/supplements/nature-index-2017-science-cities/science-and-technology-commission-of-shanghai-municipality.pdf

Aguila, M. V., Garrido Yserte, R., & Gallo Rivera, M. T. (2010). *Los sistemas regionales de innovación: una caracterización para el caso de Chile*. International Meeting on Regional Science – AECR, 1–32.

Alfonso, W. (2014). Urbanization: Concepts, Trends and Analysis in Three Latin American Cities. *Miscellanea Geographica – Regional Studies on Development, 18*(3), 5–15.

Aritenang, A. F. (2013). The Role of Technology in Regional Development: TFP and Econometrics Analysis. *Journal of Indonesian Economy and Business, 28*(1), 149–158.

Bogotá Government. (2016). *Preliminary Project of Development Plan 2016–2019*. Retrieved from http://aplicaciones.sdp.gov.co:7777/pdd/anteproyecto_plan_distrital_desarrollo_2016_2019.pdf

Braconier, H. (2000). Do Higher Per Capita Incomes Lead to More R&D Expenditure? *Review of Development Economics, 4*(3), 244–257.

Bronzini, R., & Piselli, P. (2009). Determinants of Long-run Regional Productivity with Geographical Spillovers: The Role of R&D, Human Capital and Public Infrastructure. *Regional Science and Urban Economics, 29*, 187–199.

Caicedo Aspril, H. (2011). El papel de los sistemas regionales de innovación en ciudades región globales. *Cuadernos de Administración, 27*(45), 95–108.

Camagni, R. (2003). Incertidumbre, capital social y desarrollo local: enseñanzas para una gobernabilidad sostenible del territorio. *Investigaciones Regionales, 2*, 31–57.

Capdevielle, M., & Dutrénit, G. (2012). Políticas para el Desarrollo productivo y la innovación: desafío y oportunidad para la economía mexicana. In J. Calva (Ed.), *Nueva estrategia de industrialización. Análisis estratégico para el desarrollo* (Vol. 7, pp. 153–185). Mexico City: Consejo Nacional Universitario.

Cartajena, Y. (2016). *Smart Cities in Latin América*. United Nations Commission on Science and Technology for Development. Retrieved from http://unctad.org/meetings/en/Presentation/CSTD_2015_ppt05_Cartajena_en.pdf

Chadha, V. (2007). An Assessment of the Regional R&D Effort in India: An Inter-State Analysis of Selected Industrially Advanced States. *Studies in Regional Science, 37*(2), 415–425.

Cimoli, M., Carlos, F. J., & Annalisa, P. (2005). *Science and Technology Policies in in Open Economies: The Case of Latin America and the Caribbean* (Vol. 165). Santiago de Chile: CEPAL – Serie desarrollo productivo.

Coe, D., Helpman, E., & Hoffmaister, A. (1997). North-south Spillovers. *Economic Journal, 107*, 134–149.

Cohen, B. (2014a). Smart City Index Master Indicators Survey (2014).

Cohen, B. (2014b). *The 8 Smartest Cities in Latin America*. Retrieved from http://www.fastcoexist.com/3022533/the-8-smartest-cities-in-latin-america

Collazos Rodríguez, J., & Londoño Martínez, H. (2014). Escalafón global de ciudades para la atracción de inversión industrial en la cuenca del Pacífico latinoamericano. *Estudios Gerenciales, 30*(131), 103–114.

CONACYT. (2016). *General Report on Science, Technology and Innovation Mexico, 2016*. Retrieved from http://www.siicyt.gob.mx/index.php/transparencia/informes-conacyt/informe-general-del-estado-de-la-ciencia-tecnologia-e-innovacion/informe-general-2016

Cooke, P., & Memedovic, O. (2003). *Strategies for Regional Innovation Systems: Learning Transfer and Application*. Policy Papers. Wien: UNIDO.

Crescenzi, R., & Rodríguez-Pose, A. (2013). Regional Knowledge Creation in China, India and the US. Lessons for European Regions. In R. Capello & C. Lenzi (Eds.), *Territorial Patterns of Innovation: An Inquiry on the Knowledge Economy in European Regions* (pp. 263–298). London: Routledge.

Crescenzi, R., & Jaax, A. (2017). Innovation in Russia: The Territorial Dimension. *Economic Geography, 93*(1), 66–88.

Dettori, B., Marrocu, E., & Paci, R. (2012). Total Factor Productivity, Intangible Assets and Spatial Dependence in the European Regions. *Regional Studies, 46*(10), 1401–1416.

ECLAC. (2016). *Science, Technology and Innovation in the Digital Economy*. Retrieved from http://repositorio.cepal.org/handle/11362/40840

Gordon, J., & McCann, P. (2005). Clusters, Innovation and Regional Development: An Analysis of Current Theories and Evidence. In C. Karlsson, B. Johansson, & R. Stough (Eds.), *Industrial Clusters and Inter-firm Networks* (pp. 29–57). Cheltenham: Edward Elgar Publishing.

Goschin, Z. (2014). R&D as an Engine of Regional Economic Growth in Romania. *Romanian Journal of Regional Science, 8*(1), 24–37.

Hudson, R. (1999). The Learning Economy, the Learning Firm and the Learning Region: A Sympathetic Critique of the Limits to Learning. *European Urban and Regional Studies, 6*(1), 59–72.

Krugman, P. (1979). A Model of Innovation, Technology Transfer and the World Distribution of Income. *Journal of Political Economy, 87*, 253–266.

Ligenzowska, J. (2015). *The Impact of Innovation on the Development of the Małopolska Region*. Research Papers of Wrocław University of Economics (No. 394), pp. 64–71.

Méndez, R., García Palomares, J., & Michelini, J. (2007). La construcción de ciudades innovadoras: transformaciones económicas recientes y desarrollo local en Getafe. *Estudios Geográficos, 68*(262), 173–203.

Mexico City Government. (2016). *Mexico City in the 2016 Smart Cities List!* Retrieved from http://en.mxcity.mx/2016/12/2016-smart-cities-list/

Mukim, M. (2012). Does Agglomeration Boost Innovation? An Econometric Evaluation. *Spatial Economic Analysis, 7*(3), 357–380.

Noronha Vaz, E., Noronha Vaz, T., & Nijkamp, P. (2013). The Spatial-Institutional Architecture of Firms' Innovative Behaviour. In K. Kourtit, P. Nijkamp, & B. Stimson (Eds.), *Applied Modelling of Regional Growth and Innovation Systems (Tentative Title)*. Berlin: Springer Verlag. (in print).

Office of the Deputy Prime Minister (ODPM). (2006). *State of the English Cities – A Research Study* (Vol. 1). London: Office of the Deputy Prime Minister.

OMPI. (2015, 17 de Septiembre de). *Organización Mundial de la Propiedad Intelectual.* Obtenido de Índice Mundial de Innovación 2015: Suiza, Reino Unido, Suecia, Países Bajos y EE.UU., en cabeza de lista. Retrieved from http://www.wipo.int/pressroom/es/articles/2015/article_0010.html

Porter, M. (2000). *Sobre competitividad.* Boston: Harvard Business Publishing.

Posner, M. (1961). International Trade and the Technical Change. *Oxford Economic Papers, 13*, 323–341.

Ramirez J., J. C., & Parra-Peña S., R. I. (2010). *Escalafón de la competitividad de los departamentos en Colombia.* Bogotá: CEPAL – Serie Estudios y Perspectivas.

Rodríguez-Pose, A., & Villareal Peralta, E. M. (2015). Innovation and Regional Growth in Mexico: 2000–2010. *Growth and Change, 46*(2), 172–195.

SEDECO. (2016). *Connectivity Plan Mexico City GCDMX.* Retrieved from http://www.sedeco.cdmx.gob.mx/comunicacion/nota/inicia-gcdmx-plan-maestro-de-conectividad

Solow, R. (1956). A Contribution to the Theory of Economic Growth. *The Quarterly Journal of Economics, 70*(1), 65–94.

UNCTAD. (2013). *Science, Technology and Innovation for Sustainable Cities and Peri-urban Communities.* Retrieved from http://unctad.org/meetings/en/SessionalDocuments/ecn162013d2_en.pdf

Usai, S. (2011). The Geography of Inventive Activities in OECD Regions. *Regional Studies, 45*(6), 711–731.

Valdez-Lafarga, C., & León-Balderrama, J. I. (2015). Hacia una taxonomía de los sistemas regionales de innovación en México. *Economía, Sociedad y Territorio, 15*(48), 517–553.

Wood, P. (2009). Service Competitiveness and Urban Innovation Policies in the UK: The Implications of the 'London Paradox'. *Regional Studies, 43*(8), 1047–1059.

CHAPTER 2

Applying a New Methodology to Measure Investment in R&D and Science and Technology Activities: The Case of Colombia

Alexander Cotte Poveda, Jorge Andrade Parra, and Clara Inés Pardo Martínez

2.1 Introduction

Investments in science, technology and innovation are important in developing countries, but they must be analyzed from the perspective of research and development (R&D) to provide different advantages such as the following: (1) increasing competitiveness and productivity because

A. Cotte Poveda (✉)
Universidad Santo Tomás, Bogotá, Colombia

Colombian Observatory of Science and Technology (OCyT), Bogotá, Colombia

J. Andrade Parra
Colombian Observatory of Science and Technology (OCyT), Bogotá, Colombia

C. I. Pardo Martínez
School of Management, Universidad del Rosario, Bogotá, Colombia
e-mail: clara.pardo@urosario.edu.co

C. I. Pardo Martínez et al. (eds.), *Analysis of Science, Technology, and Innovation in Emerging Economies*,
https://doi.org/10.1007/978-3-030-13578-2_2

R&D can strengthen domestic industry and agriculture by adding value to local products and services; (2) supporting social issues because R&D can be used to address societal challenges such as the water and energy supply, health, food security, climate change, etc. through solutions based on knowledge and new technologies; (3) developing new technology to resolve local problems; (4) generating new knowledge to improve the capacity and skills of human resources; (5) enhancing the scientific and intellectual community through investments in R&D, higher education and the creation and retention of researchers to resolve productive and social problems based on science and technology; (6) diversifying the economy toward more value-added activities and structural changes that are drivers of economic prosperity (Bell & Albu, 1999; Cohen & Levinthal, 1990; Hausmann & Hidalgo, 2011).

In this context, government interventions are important for balancing low levels of private investment in R&D and for addressing and solving societal challenges, for which policies should take account for market failures through subsidies for R&D (Arrow, 1962), systematic failures that might slow down or even block interactive learning, R&D and innovation (Smith, 2000) and public-value failures that are generated when neither the market nor public sector provides goods and services required to achieve public values (Bozeman & Sarewitz, 2011). Therefore, it is necessary to analyze trends in the main indicators related to expenditure in R&D to provide a compelling rationale for and to evaluate different policy instruments to encourage the maintenance and expansion of R&D investments given competing development agendas, social problems and challenges, and severe financial constraints.

Sustainable development and economic growth are the result of innovations, a skilled labor force, research productivity, growth in R&D expenditure, high-quality institutions, technology diffusion and human capital, among other factors. Table 2.1 presents different studies that have demonstrated the relationship between R&D and economic growth and development.

Other studies have analyzed alternative measures to examine R&D. For example, Salimi and Rezaei (2018) measured different levels of R&D performance by applying a multi-criteria decision-making approach named the Best Worst Method to determine the importance of R&D to 50 high-tech Small and Medium Enterprises in the Netherlands, and they established the importance of different perspectives in formulating more effective strategies to improve R&D in accordance with priorities. Gault

Table 2.1 Main studies of the relationship between R&D and economic growth and development

Study	*Main implications*
Proposed model to analyze how the long-term economic growth of an economy depends on globally developed ideas (Jones, 2002).	The main sources of economic growth in the USA include education, R&D and global population growth.
The contribution of R&D and human capital to the Australian economy using country-level data during the period of 1960–2000 (Chou, 2002).	The growth of Australian per capita income is due to both factor accumulation and enhancing the efficiency of transforming inputs into outputs, in which R&D plays an important role.
Korean growth experience from the neoclassical growth perspective (Kwack & Lee, 2006).	Investment rate, R&D, education and the size of government were the major determinants of long-term economic growth in Korea. Moreover, public and household expenditures on education and R&D investment are major contributors to innovation and improving the quality of labor.
The contribution of R&D subsidies and publicly provided science education to economic growth (Grossmann, 2007).	It was determined that public education contributes to economic growth more efficiently than R&D subsidies, so it is important to develop R&D by promoting the public education of scientists and other skilled people.
Role of R&D investment in the long-term economic growth of OECD countries using a panel data set (Falk, 2007).	This study found that higher R&D investment was positively related to GDP growth in working-age populations.
The trends in various components of R&D and its contribution to economic growth in the USA (Goel, Payne, & Ram, 2008).	It established a strong association between economic growth and federal R&D expenditure rather than non-federal R&D outlays. Moreover, R&D is key for sustainable economic growth.
The association of regional disparities in R&D and higher education with regional economic growth in Europe (Sterlacchini, 2008).	It demonstrated a positive and significant effect of knowledge, educational attainment and intensity of R&D expenditure on the economic growth of European regions, where innovation and knowledge, public support for higher education and R&D have been an effective tool to promote growth and development.

(*continued*)

Table 2.1 (continued)

Study	*Main implications*
The causal relationship between increased research productivity and economic growth in five Asian countries using the Granger causality framework (Jin, 2009).	Offering numerous types of services and tertiary education may have a direct and immediate effect on services in management and other sectors, and research productivity causes economic growth.

(2018) analyzed different strategies to measure innovation and proposed a taxonomy and framework to account for supporting policy development, examining and evaluating implemented policies, and policy-related learning resulting from analysis. Aksnes, Sivertsen, Leeuwen, and Wendt (2017) evaluated methodological problems in measuring research productivity at the national level by comparing official R&D statistics from the OECD using different indicators.

These studies demonstrated the importance of using adequate indicators to measure R&D expenditures and scientific, technological and innovation activities to inform adequate science, and technology and innovation (STI) policies that satisfy the requirements and address the challenges of the country.

The Frascati Manual is a set of guidelines for collecting and reporting data on research and experimental development that is internationally comparable, and such analysis is a fundamental tool for the institutions responsible for these calculations and policy-makers to determine reliable, robust and clear indicators to analyze trends in R&D (OECD, 2015). The main changes in this version of the Frascati Manual are as follows: (1) updates to the main concepts and definitions and the undertaking of new approaches to finance and execute R&D and scientific, technological and innovation activities (STIA); (2) improved international comparability of R&D and STIA indicators; (3) generation of a set of measurement guidelines with relevant and applicable characteristics for emerging economies; (4) facilitation of homologation of R&D and STIA data from the Frascati Manual and the national accounts of countries; (5) standardization of the Frascati Manual with other classification and statistical systems of complementary manuals related to science, technology and innovation known as the "Frascati family"; (6) promotion of better R&D and STIA measurement practices to facilitate the comparability of information; (7) emphasis on the overriding importance of administrative records to measure R&D and STIA indicators that guarantee higher reliability, robustness and transparency than traditional survey approaches; and (8) consideration of the use of R&D micro-data, their connections and relationships to potentially develop analysis. These changes imply the application of new methods and measurement strategies to guarantee that R&D indicators account for the analysis of private and public accountability systems to identify specific budgets related to science, technology and innovation according to the guidelines of the Frascati Manual.

With this background, the objective of this chapter is to describe the methods used to update and apply the measurement of expenditure in R&D and other scientific, technological and innovation activities following the guidelines of the Frascati Manual to develop primary indicators in the Colombian context based on the work by the Colombian Observatory of Science and Technology (OCyT), which is the entity responsible for calculating these indicators every year at different levels of aggregation.

This chapter describes the calculation and analysis of the indicators for every sector suggested by Frascati, that is, business enterprise, government, higher education and private non-profit, considering mixed methods using administrative registers and surveys as well as validations with other sources.

This chapter is organized as follows. The first section shows a background and importance of this study. The second section describes the main methods used in this study following the main requirements of the Frascati Manual. The third describes the main results of application of the Frascati Manual in the Colombian context for different sectors. Finally, the main reflective conclusions that emerged from this study as well as suggestions for future work are discussed.

2.2 Data and Methods

The study is based on the Frascati Manual (2015) for the 2000–2017 sample period with a projection to 2018. Data were recollected from different sources, and annual surveys were developed by OCyT for the sectors proposed by the manual as follows.

Measuring Government R&D is fundamental due to the importance of these institutions as performers and funders of STIA, and when it is used, the method denominates government budget allocations for R&D based on the financing sources of government entities by identifying all budget items that support STIA. In Colombia, measurements are generated from administrative and budget registers of government institutions provided by the Colombian Ministry of Finance and Public Credit from the General Budget of the Nation (with a scope of 194 institutions for 2017), thus enabling data at any stage of the budget cycle to be analyzed, including effectively executed expenditures by every government entity that engaged in R&D activities and other STIA.

Measuring Higher Education R&D is important because of the relevant sector to the investment and promotion of STI and the design of specific policies to promote R&D and generate new knowledge. A mixed method is used that follows the Higher Education Expenditure on Research and Development and includes the survey developed by OCyT each year, as well as information from administrative registers of the executed budgets of institutions of higher education that are collected by the Colombian Ministry of National Education with a scope of 128 institutions for 2017.

Measuring Business Enterprise R&D is key to improving productivity, competitiveness and value addition from R&D activities. This measurement uses a mixed method according to the Business Enterprises Expenditure on Research and Development from a survey of development and technological innovation by the National Administrative Department of Statistics (DANE) of Colombia and financial statements provided by the Business Superintendence, while accounting for the classification of economic business activities and industrial orientation.

Measuring Health R&D was developed from a survey conducted by OCyT every year and analysis of administrative registers collected by the National Health Authority to determine the expenditures in R&D and other STIA. This sector is key to generating new knowledge in this field of research, and the scope included 21 hospitals and clinics for 2017.

Measuring Private Non-Profit R&D includes research centers recognized by Colciencias that were valid as of March 2018, non-profit organizations and associations, and private non-profit organizations serving businesses, and estimates were made using the method named Private Non-Profit Expenditure on Research and Development according to the Frascati Manual (2015) and included a total of 140 organizations for 2017 (79 research and technological development centers, 26 professional associations and NGOs, and 35 private non-profit organizations serving businesses).

Applying these methods to calculate expenditures in R&D and other STIA is important to analyze and establish adequate policies that can encourage these investments as a key element in promoting economic growth and development, especially in productive sectors. However, reliable, transparent and robust indicators for each sector must be developed using different approaches following the guidelines of the Frascati Manual.

2.3 Results

This section presents the main results for indicators related to R&D and other STIA expenditures by different sectors following the Frascati Manual (2015). All monetary indicators are expressed in 2014 Colombian pesos and account for the base year of national accounts calculated by DANE. The indicators show trends in expenditures as a percentage of the GDP, funding by resource according to the type of institution, distribution by the type of activity, and governmental funding by socioeconomic objective.

2.3.1 Trends in STIA Expenditure in Colombia

Beginning in 2011, Colombia began to increase investments in STIA and R&D; 2013 and 2014 were the years of the highest investment of 0.75% in STIA and 0.31% in R&D, mainly due to found royalties exclusively designated for STI. In recent years, the trends in these investments have been maintained (see Fig. 2.1).

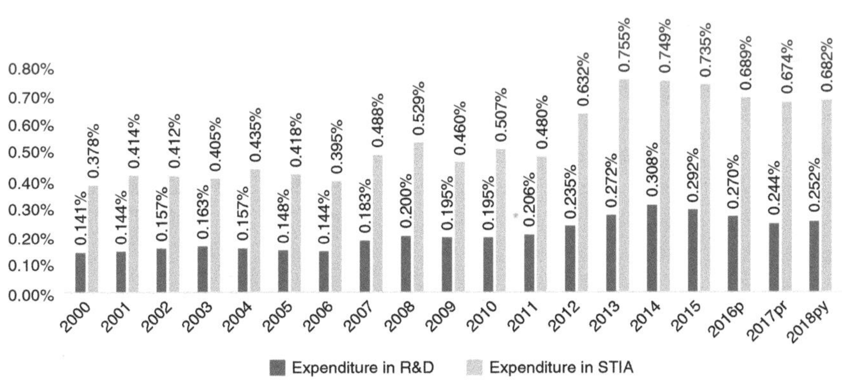

Fig. 2.1 Trends in STIA as a percentage of GDP. Source: OCyT (2018). Note: *p* provisional data, *pr* preliminary data, *py* projected data

From 2013 to 2017, public resources decreased, generating higher investment in STIA by the private sector of 37.43% and 62.35% on average, respectively, and the same situation applied to R&D with 30.57% and 68.98% on average, respectively. These results indicate the importance of maintaining private investments as well as the need to increase public investment, especially for research priorities in which the private sector has no interest, so public investment is important for resolving key problems for the country. Private R&D is often related to the ability of a private firm to access data, information, technologies, materials, risks and capital from foreign sources, so the relative openness of a given economy can determine the degree of private investment (Gisselquist & van der Meer, 2001).

2.3.2 *Investments in STIA by Colombian Institutions*

In recent years, investments in STIA by different institutions have varied. Firms, government organizations and institutions of higher education made the greatest investments of 53.39%, 20.76% and 16.65% of the total, respectively, in 2017. Table 2.2 shows the main trends in investments by institutions.

Maintaining or increasing investment in STIA, especially in R&D, is critical to the production of new technologies to enhance productivity and reduce poverty in many developing and emerging economies. In these cases, the potential of the private sector has yet to be realized, particularly with respect to the production of technologies that are explicitly relevant to small-scale improvement in resource use and that support vulnerable social groups, for which it is important to improve or design adequate instruments to promote investment from both private and public sectors to generate knowledge according to the requirements of the country. With the right combination of elements and features to create an enabling environment and appropriate incentive mechanisms, the question becomes how institutions respond in terms of investing in STIA and R&D (Naseem, Spielman, & Were, 2010).

2.3.3 *Distribution of STIA Expenditure by Type of Activity*

STIA includes the following components: R&D, support for scientific and technological training, administration, and other supporting and innovation activities that are key to generating new knowledge and technological development. In Colombia, the main STIA activities are R&D and innovation activities (see Fig. 2.2).

Table 2.2 The main trends in STIA investments by institutions, 2009–2017

Institution	*2009*	*2010*	*2011*	*2012*	*2013*	*2014*	*2015*	*2016p*	*2017pr*
Firms	29.69%	30.61%	28.91%	37.08%	31.18%	41.99%	46.84%	50.87%	53.39%
Higher education institutions	27.06%	22.68%	23.33%	19.06%	19.17%	19.19%	17.20%	16.65%	13.02%
Government organizations	24.48%	30.22%	27.95%	29.67%	36.66%	26.59%	23.56%	19.94%	20.76%
Research and technological development centers	12.68%	10.61%	13.97%	10.09%	9.71%	8.77%	9.07%	9.15%	9.49%
Hospitals and clinics	2.64%	2.51%	2.54%	1.78%	1.45%	1.47%	1.36%	1.38%	1.36%
Private non-profit organizations serving businesses	2.71%	2.66%	2.51%	1.78%	1.38%	1.45%	1.53%	1.52%	1.53%
Professional associations and NGOs	0.73%	0.71%	0.78%	0.53%	0.45%	0.54%	0.43%	0.49%	0.45%

Source: OCyT (2018)

Note: *p* provisional data, *pr* preliminary data

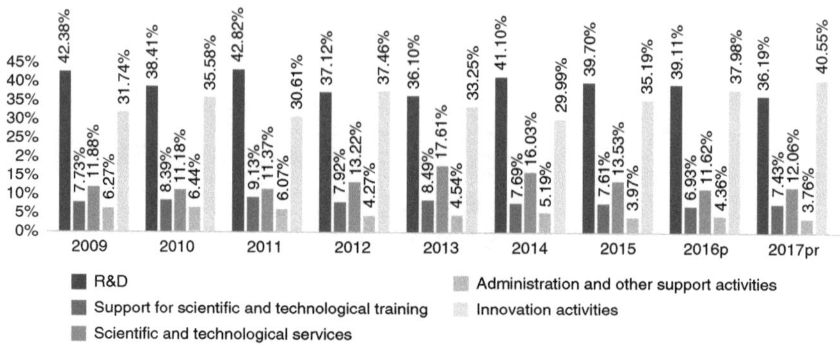

Fig. 2.2 Trends in STIA investment as a percentage of GDP. Source: OCyT (2018). Note: *p* provisional data, *pr* preliminary data, *py* projected data

According to the G7 Academies of Science (2017), governments should lead the following aspects: (1) expanding investments and capacity in science and pre-competitive technologies and innovations; (2) increasing investments in STI infrastructure (tangible and intangible) that support inclusive development and scientific progress; (3) promoting the development of capacities to design, engineer, produce and deliver goods and services based on new knowledge and technology; (4) promoting open science while accounting for intellectual property to prevent the emergence of monopolistic practices; (5) sharing effective practices and process to develop policies and instruments that promote and enhance innovation, technological and scientific dissemination, and efficient infrastructure generation with the aim of integrating countries with different levels of development; and (6) ensuring that an appropriate governance framework that benefits STI be fully developed and implemented while maintaining the public trust. These are important elements in the design and application of adequate STI policies, especially in emerging economies.

2.3.4 *Governmental STIA Funding by Socioeconomic Objective*

Another important point related to STIA is how public resources are distributed according to socioeconomic objective. In applying the Frascati Manual (2015) in Colombia, it was found that the investment in R&D

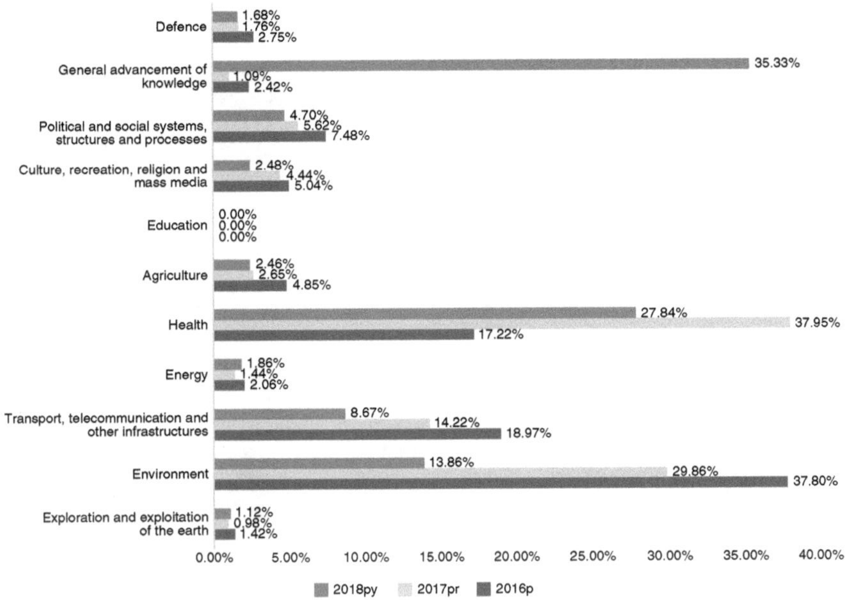

Fig. 2.3 Investments in R&D by socioeconomic objective. Source: OCyT (2018). Note: *p* provisional data, *pr* preliminary data, *py* projected data

was predominantly in health and the environment in 2017 (see Fig. 2.3), whereas the main investments in STIA were toward the general advancement of knowledge and education (see Fig. 2.4).

STI is important for achieving socioeconomic objectives, so the Colombian government must prioritize the promotion of regional parity and knowledge according to the different contexts and requirements of regions to achieve sustainable development throughout the whole country. Additionally, knowledge, technology and innovation must be promoted as a fundamental pillar of the new society based on the appropriate use of STI in pursuit of human welfare and equality, so it is fundamentally necessary to measure and maintain adequate STIA indicators as inputs for policy and decision makers based on the international guidelines of the Frascati Manual (2015).

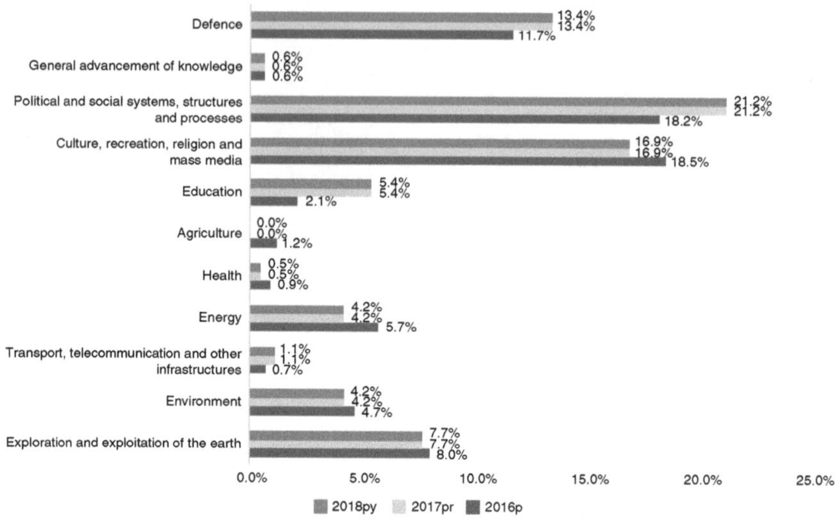

Fig. 2.4 Investments in STIA by socioeconomic objective. Source: OCyT (2018). Note: *p* provisional data, *pr* preliminary data, *py* projected data

2.4 Conclusions

This chapter presents the application of the Frascati Manual (2015) for Colombia with an update for the period of 2000 to 2017 for the sectors of government, higher education, business enterprise, health and non-profit organizations, and accounts for the main indicators of expenditure and investment in R&D and other STIA.

The methods used in this study have multiple implications and advantages such as improving the comparability of R&D expenditure according to international standards (OECD, Eurostat), providing information on actual budget execution from administrative and accounting registers related to STI, reviewing historical data from the period 2000–2017 for R&D and other STIA investments at different levels of aggregation, identifying stable and high-quality indicators that can be verified from administrative registers, optimizing the time spent collecting data and information, clarifying the relationship between STIA and public policy as classified by socioeconomic objective, analyzing different sources of finance that include international organizations, and identifying reproducible indicators that guarantee adherence to international guidelines, which implies their robustness, reliability and statistical usefulness.

This study covered 100% of the government institutions that are governed by the general budget at the national level that was executed on December 31, 2017. The methods included the General System of Royalties, updated all the time-series data from 2000 to 2017, and analyzed and verified the results of surveys of administrative registers from different sources that discriminate expenditures and investments in STI from accounting and financial statements, among other sources.

It is important to continue applying the guidelines of the Frascati Manual considering regional governments, to increase the use of administrative registers for business and higher education as data sources, to only survey for information that cannot be captured from administrative registers and to strengthen the importance of adequate STIA registers in the accounting systems of different private institutions such as universities, enterprises and non-profit institutions, and to incorporate the International Financial Reporting Standards into their respective accounting records. This will enable better records of STI expenditures and investments in the country.

These indicators are an input that can explain different market and institutional factors as well as trends in and the dynamics of STIA investments, for which multiple incentive mechanisms, strategies and programs would stimulate investment in different sectors. Push-and-pull mechanisms that stimulate both the demand for and supply of private STIA and R&D—if cautiously designed, effectively funded and politically supported—could produce desirable stimulus effects. However, it remains to be seen which processes are most effective and under what circumstances. Further research is needed to develop new strategies to implement these incentive mechanisms and assess their impact on STIA investment.

References

Aksnes, D., Sivertsen, G., Leeuwen, D., & Wendt, K. (2017). Measuring the Productivity of National R&D Systems: Challenges in Cross-national Comparisons of R&D Input and Publication Output Indicators. *Science and Public Policy, 44*, 246–258.

Arrow, K. (1962). Economic Welfare and the Allocation of Resources for Invention. In Universities-National Bureau Committee for Economic Research, Committee on Economic Growth of the Social Science Research Council (Ed.),

The Rate and Direction of Inventive Activity. Princeton, NJ: Princeton University Press.

Bell, M., & Albu, M. (1999). Knowledge Systems and Technological Dynamism in Industrial Clusters in Developing Countries. *World Development, 27*, 1715–1734.

Bozeman, B., & Sarewitz, D. (2011). Public Value Mapping and Science Policy Evaluation. *Minerva, 49*, 1–23.

Cohen, W., & Levinthal, D. (1990). Absorptive Capacity: A New Perspective on Learning and Innovation. *Administrative Science Quarterly, 35*, 128–152.

Chou, Y. K. (2002). *The Australian Growth Experience (1960–2000), R&D Based, Human Capital-based or Just Steady State Growth?* Research Paper No. 855. Department of Economics, University of Melbourne.

Falk, M. (2007). R&D Spending in the High-tech Sector and Economic Growth. *Research in Economics, 61*, 140–147.

G7 Academies of Science. (2017). *New Economic Growth: The Role of Science, Technology, Innovation and Infrastructure*. G7 Academies' Joint Statements 2017. Retrieved from https://royalsociety.org/~/media/about-us/international/g-science-statements/2017-may-3-new-economic-growth.pdf?la=en-GB

Gault, F. (2018). Defining and Measuring Innovation in All Sectors of the Economy. *Research Policy, 47*, 617–622.

Gisselquist, D., & van der Meer, C. (2001). Regulations for Seed and Fertilizer Markets: A Good Policy Guide for Policy Makers (Rural Development Working Paper No. 22817). Washington, DC: World Bank.

Goel, R. K., Payne, J. E., & Ram, R. (2008). R&D Expenditures and U.S. Economic Growth: A Disaggregated Approach. *Journal of Policy Modeling, 30*, 237–250.

Grossmann, V. (2007). How to Promote R&D-based Growth? Public Education Expenditure on Scientists and Engineers Versus R&D Subsidies. *Journal of Macroeconomics, 29*, 891–911.

Hausmann, R., & Hidalgo, C. (2011). The Network Structure of Economic Output. *Journal of Economic Growth, 16*, 309–342.

Jin, J. C. (2009). Economic Research and Economic Growth: Evidence from East Asian Economies. *Journal of Asian Economics, 20*, 150–155.

Jones, C. I. (2002). Sources of U.S. Economic Growth in a World of Ideas. *The American Economic Review, 92*, 220–239.

Kwack, S. U., & Lee, Y. S. (2006). Analyzing the Korea's Growth Experience: The Application of R&D and Human Capital Based Growth Models with Demography. *Journal of Asian Economics, 17*, 818–831.

Naseem, A., Spielman, D., & Were, S. (2010). Private-Sector Investment in R&D: A Review of Policy Options to Promote its Growth in Developing-Country Agriculture. *Agribusiness, 26*(1), 143–173.

OCyT. (2018). *Indicators of Science and Technology Colombia*. Bogotá, Colombia.

OECD. (2015). *Frascati Manual*. Retrieved from http://www.oecd.org/publications/frascati-manual-2015-9789264239012-en.htm

Salimi, N., & Rezaei, J. (2018). Evaluating Firms' R & D Performance Using Best Worst Method. *Evaluation and Program Planning, 66*, 145–157.

Smith, K. (2000). Innovation as a Systemic Phenomenon: Rethinking the Role of Policy. *Enterprise and Innovation Management Studies, 1*, 73–102.

Sterlacchini, A. (2008). R&D, Higher Education and Regional Growth: Uneven Linkages Among European Regions. *Research Policy, 37*, 1096–1107.

CHAPTER 3

Bioprospecting Model for a New Colombia Drug Discovery Initiative in the Pharmaceutical Industry

Juan Bueno and Sebastian Ritoré

3.1 Introduction

The natural sciences and biotechnology are currently recognized worldwide as the trends in the knowledge-based economy, after information technology, that create the most new opportunities for societies and economies (Pyka & Prettner, 2018). It is therefore important to take into account the fact that there are two types of impact for policies relating to new ventures: the first is direct impact, which refers to the use of the products that are generated by the activity, and the second is indirect impact, which refers to the appropriation of the products and the benefit obtained from them (Schaffartzik et al., 2015). It is considered that the development of new medicines is one of the most promising results of biotechnological initiatives (Kamuriwo & Baden-Fuller, 2016). However, one of the most important weaknesses identified in the multidisciplinary process

J. Bueno (✉) • S. Ritoré
Research Center of Bioprospecting and Biotechnology for Biodiversity Foundation (BIOLABB), Bogotá, Colombia
e-mail: juanbueno@biolabb.org

C. I. Pardo Martínez et al. (eds.), *Analysis of Science, Technology, and Innovation in Emerging Economies*,
https://doi.org/10.1007/978-3-030-13578-2_3

applied to drug discovery is the lack of experience in negotiation and in the managerial skills and abilities of scientific researchers. In addition, there are still many opportunities for improving legislation regarding the management and rights over genetic resources and intellectual property, as well as in the promotion of market analysts in the field of bioprospecting (Goyes & South, 2015; Hernández, 2008). Taking all this into account, any developing country that decides to become involved in this area must consider the fact that modern biotechnology has certain distinctive characteristics (De Witt, Osseweijer, & Pierce, 2017):

- There is a close relationship between science and business.
- The companies that are involved in biotechnology invest a high percentage of their budget in research and development (R&D); they are knowledge companies.
- Genetically modified products must jump through particular hoops to reach the market (Gonzales, Villa, & Velasco, 2007).

To develop these objectives, bioprospecting and biotechnology have come together as related disciplines to evaluate the sustainable use of biodiversity, determining the components as well as the properties of biological and genetic resources, and to produce them industrially, avoiding non-sustainable extraction (Gallón, 2011; Pushpangadan, Ijinu, Dan, & George, 2015). Similarly, innovation in biotechnology, both in products and processes, has a positive impact on the industry, with the chemical industry being the biggest beneficiary of this technology transfer (Jr Silva, Silveira, Paranhos, Hasenclever, & Miranda, 2017; Zanello, Fu, Mohnen, & Ventresca, 2016).

In Colombia, despite the fact that bioprospecting is important, this very broad activity still does not have a specific focus on products that are of industrial or commercial interest (Prip & Rosendal, 2015; Wyatt & Brisman, 2017). This results from an absence of scientific governance, which could guide bioprospecting activity towards the search for and identification and isolation of genes, proteins, and secondary metabolites, as well as the evaluation of their potential use for medicine, agriculture, and industry (Cotes Prado et al., 2012; Departamento Nacional de Planeación, 2011). In addition, these activities are limited further because the required information either is dispersed or does not exist. The bioprospecting projects that have been carried out only seek publication, with no attempt being made to control information about the taxonomic, morphological, and physiological description of the different species or to establish their economic possibilities (Departamento Nacional de Planeación, 2011; Rodríguez et al., 2009). Moreover, the absence of a science that uses the results obtained from bioprospecting to develop applications that solve the problems of populations and ecosystems that are increasing the loss of biodiversity is a constant threat for countries that require diversity as a model of growth (Purkayastha, 2016). Therefore, it is a priority for Colombia to develop the sector that includes biotechnology-based companies, as they provide added value and allow the sustainable use of biodiversity: this is a great opportunity for innovation in the productive sector (Jiménez, Cárdenas, & Soler-Tovar, 2017; Wyatt & Brisman, 2017). These companies are the bridge between the results of research and the application of these results in production systems that use the market. For this development to take place requires a favorable environment for the commercial development of biotechnology and the sustainable commercial use of biodiversity (Vargas-Hernández, Pallagst, & Hammer, 2017). The objective of this chapter is to offer a rational bioprospecting model that can support a biotechnological industry and also allows businesses to grow (Bueno, 2017a).

3.2 The Biodiversity of Colombia

First, it is necessary to quantify the biotechnological potential of biodiversity in Colombia as it relates to the sustainable production of pharmaceutically relevant compounds: this includes an analysis of bioprocess design, bioprocess technology, footprint, and production process economy (Kusari, Singh, & Jayabaskaran, 2014). Colombia takes up 0.22% of the earth's surface and has more than 10% of all known species, these organisms coexisting in 311 types of continental and coastal ecosystems. It is among the 14 countries with the greatest biodiversity on the planet, with this large number of species being distributed in five ecoregions: Pacific region, Caribbean, Amazon, Orinoquia, and Andean (Moreno, Andrade, & Ruíz-Contreras, 2016). So far, 25,163 plant species, 6222 vertebrates, 70,082 invertebrates, 1766 fungi, and 1464 lichens have been discovered in the country (Moreno et al., 2016), which represents a great source of biomass for a rational model of bioprospecting that is looking for secondary metabolites that might have a considerable impact in medicine and industry. Among the most representative ecosystems is the *páramo*, an ecosystem that is typical of tropical mountains located above

Picture 1 The *páramo* ecosystem represents 17% of Colombia's floral diversity, yet covers only 2.5% of the continental territory (Marín & Parra, 2015)

the upper limit of forests, with its own flora that reaches 60% of plants and has 40% of endemic vegetal species (4700 species of plants) in its biogeographic zone, the Andean region (see Picture 1) (Marín & Parra, 2015; Rangel, 2015).

3.3 Sustainable Biocommerce: The Valuation of Life and Application of Biodiversity to Benefit the Prosperity of Colombia

The third Conference of the Parts of the Agreement of Biological Diversity created and adopted trade initiatives for green growth under the term biocommerce (Mendoza, 2014). This is defined as the use and implementation of business innovations based on available biological knowledge, and is one of the great strategies for sustainable development and for obtaining patentable products from biodiversity under a model that shares benefits between communities, industry, and academic institutions (see Picture 2) (Quezada, 2003). Similarly, the Organisation for Economic

Picture 2 Any biotrade initiative must be registered under the Nagoya protocol, so that the benefits are shared equitably in the communities affected (Kang, Jung, & Ryu, 2015)

Co-operation and Development (OECD) defines biotechnology as "The application of science and technology to living organisms, as well as parts, products and models thereof, to alter living or non-living materials for the production of knowledge, goods and services" (Montero, 2011).

Taking this into account, any business idea that seeks to add value to the organisms present in the different ecosystems requires a rational bioprospecting program that certifies the uses of biodiversity (medicine, agriculture, and industry) and protects the knowledge obtained in order to conserve the natural resources studied (Srivastava, 2016). Likewise, any bioprospecting effort must be regulated under the Convention on Biological Diversity (CBD) and the Nagoya protocol, to avoid biopiracy and ensure the fair and equitable distribution of the benefits of resource use (Departamento Administrativo de Ciencia, Tecnología e Innovación-COLCIENCIAS, 2017). In addition, this strategy of entrepreneurship for green businesses is directed as a priority towards the promotion of companies that are involved in issues such as the efficient use of energy, adaptation to climate change, waste management, cleaner technologies, sustainable building materials, and sustainable aspects of the biodiversity, biotechnology, and agro-industries (Mazzucato, 2015). Furthermore, the national green business plan for Colombia is formulated as a vision for the year 2025, by which time green businesses will be positioned and consolidated as a new strategic component of the national economy. As part of this, eight strategic lines of intervention are proposed (communication, positioning, and awareness for the consumer and producer about green businesses, policy, and regulations; science, technology and innovation; resources, and economic and financial incentives; market access, both the coordination of this and how it relates to institutions; sectoral coordination; a market information system; monitoring and evaluation; and development and strengthening of the offer). This is in order to comply with the objectives set, the mission, the vision of green business, and the goals of the production and sustainable consumption policy.

Bioprospecting represents for Colombia an alternative route for economic development, generation of employment, conservation of biodiversity, training of highly qualified people, organization and alliances between national and international research groups, and strengthening of the country's technological platform and associated production chains in some industrial sectors. In this context, a national program for sustainable agriculture and bioprospecting has been proposed with a focus on sustainability for obtaining new value chains. In the areas of agricultural biotechnology and biochemistry, owing to the sector's demands, specialists with emphasis on metabolite bioprospecting, production of bioproducts, among others (Ministerio de Agricultura y Desarrollo Rural, 2017).

The National Planning Department of Colombia has identified that the major limitations for the commercial development of biotechnology, especially for the promotion of investments in the sector, are related to the following aspects:

1. The regulation of access regulations and the lack of methodologies to assign the fair and equitable distribution of benefits. Currently the country has not been able to finalize the prior consultation for the regulation of Andean Decision 391 on access to genetic resources.
2. The low capacity for the development of bioprospecting activities in the country. There are approximately 20 modern bioprospecting centers in the country, while Brazil has more than 800 and Costa Rica has 43.

3. The minimal development of biotechnology-based companies. Colombia has a smaller number of companies dedicated to biotechnology when compared with other countries (3.2 biotechnology-based companies per million inhabitants) (https://www.dnp.gov.co/atencion-al-ciudadano/Paginas/preguntas-frecuentes.aspx).

Because of these factors, the National Planning Department has established a strategy for green growth in the National Development Plan 2014–2018 "All for a new country" (Colombia, Law 1753 of 2015). Article 156 reads: "The National Government, Through the National Planning Department, in coordination with the Ministry of Environment and Sustainable Development will define a long-term green growth policy in which the objectives and goals of sustainable economic growth are defined. Within its strategies, a program to promote research, technological development and innovation will be designed to strengthen national and regional competitiveness based on products and activities that contribute to sustainable development and contribute to green growth."

Unfortunately, the lack of rational initiatives in bioprospecting has allowed the appearance in Colombia of the phenomenon of macro-biopiracy. This has been further aggravated by the marginalization and displacement of communities, as well as the loss of traditional knowledge and the loss of tenure of the earth, all of which has led to a corresponding loss of biological resources (Goyes & South, 2015).

The situation reflects an economy in which R&D that requires an impulse to innovation is scarce, not in terms of discourse but in the obtaining and developing of products. This is reflected in a low competitiveness index (Consejo Privado de Competitividad, 2016).

3.4 Problem Tree Analysis: To Understand the Problem Is to Have the Solution

An analysis of the bioprospecting industry and its inability to obtain new products from biodiversity may be based on a problem tree. This has the following characteristics (Domínguez Molina, 2012; Pan, 2006) (see Fig. 3.1).

Problem: absence of an institutional capacity in Colombia for the commercial development of biotechnology based on the sustainable use of biodiversity.

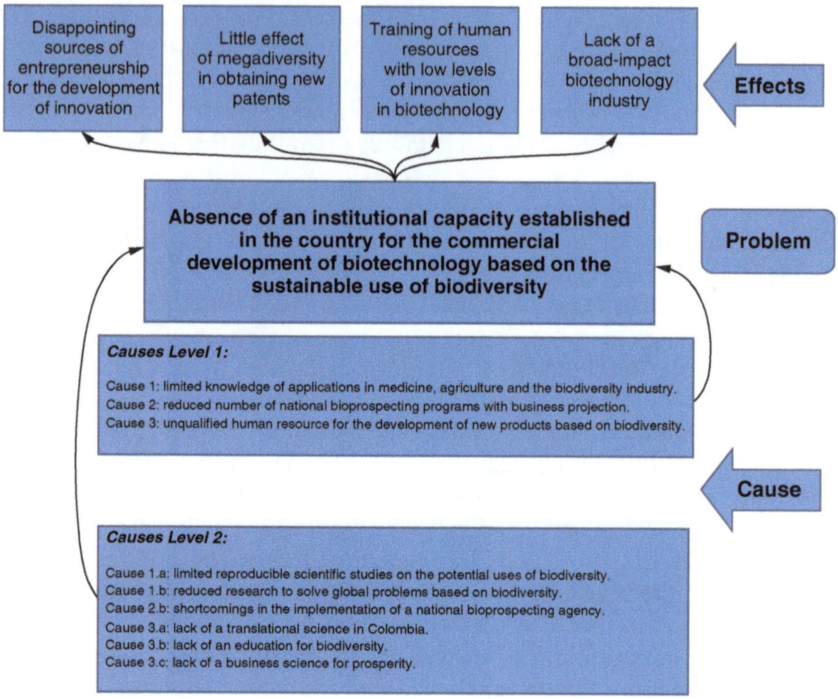

Fig. 3.1 Problem tree analysis of bioprospecting and the biotechnology industry in Colombia

Causes, Level 1

Cause 1: limited knowledge of applications related to medicine, agriculture, and the biodiversity industry.
Cause 2: reduced number of national bioprospecting programs with a business projection.
Cause 3: personnel who are unqualified for the development of new products based on biodiversity.

Causes, Level 2

Cause 1a: limited reproducible scientific studies on the potential uses of biodiversity.
Cause 1b: reduced research for solving global problems based on biodiversity.

Cause 2a: shortcomings in the implementation of a national bioprospecting agency.
Cause 3a: lack of translational science, that can apply the discoveries made by a basic science.
Cause 3b: lack of education about biodiversity.
Cause 3c: lack of a business science relating to prosperity.

Effects

1. Disappointing sources of entrepreneurship for the development of innovation.
2. Little effect of mega-diversity in obtaining new patents.
3. Training of a workforce with low levels of innovation in biotechnology.
4. Lack of a broad-impact biotechnology industry.

3.5 SWOT Analysis for Implementation of a Bioprospecting Program in Colombia

In view of the foregoing, in order to establish an institutional capacity that seeks the commercial development of biotechnology, based on the sustainable use of biodiversity, the following SWOT matrix (strengths, weaknesses, opportunities, and threats) has been designed, to inform the strategic plan that is required (see Table 3.1).

3.5.1 Market Study on the Biopharmaceutical Industry in Colombia

The structure of the biotechnology market industry comprises three types of organizations: small start-ups, medium-sized companies, and large multinationals (Moscoso, Estrada, Diaz, & Andrade, 2015). In Colombia, 73% of the market is represented by small local companies with around 50 employees by 50% average size, followed by a 28% average size and only 30 laboratories (22% of the country's total) have more than 200 employees, who generate 73% of the jobs in the sector and contribute to 83% of total production, currently the multinationals control the Colombian market (Moscoso et al., 2015).

Following this line of thinking, the global OECD report on the bioeconomy until 2030 designs a political agenda in which biotechnology will impact on three important fields: medicine, agriculture, and industry

Table 3.1 SWOT analysis for bioprospecting and biotechnology enterprises in Colombia

	Strengths	*Weaknesses*
	• Megadiversity • Access to the genetic resource • Strategic situation	• Little knowledge • No unqualified staff • Local governments
Opportunities • Biobusiness • Biotechnology • Nanotechnology	*eStrategies SO* • Implementing biodiversity Bioprospecting research centers • Biotechnology of biodiversity • Increase in the added value of biodiversity	*eStrategies WO* • Increase in the number of patents based on biodiversity • Training of human resources • Commercialization of patents and royalties
Threats • Biopiracy • Good scientific practice • Little assessment	*eStrategies ST* • Legislation on biopiracy • Legislation on scientific misconduct • Biodiversity communication strategy as a source of development	*eStrategies WT* • Science as enterprise—knowledge as added value • Biodiversity and its conservation as heritage

(Arundel & Sawaya, 2009). It is possible, therefore, to observe the great opportunity that will be presented to the country if a rational program of bioprospecting is implemented, one that seeks to commercialize products coming from biodiversity and lead Colombia towards a sustainable bio-economy. This national plan of applied research will be capable of developing products and services that give added value to the evaluation and conservation of our ecosystems.

The drug market depends on the discovery of new chemical entities (NCEs) with biological activity. These are classified into 14 main categories, which compare the chemical structure with the pharmacology and the organ where the drug acts. No net quantity (NEQ) is included in the classification if does not have a marketing request from at least one country and thus obtain United States Adopted Name (USAN) or British Approved Name (BAN) registration (registration available at http://www.whocc.no/atc/structure_and_principles/) (Cheng, Zhao, Xiao, & Chou, 2017). Likewise, the NCEs, when approved by the Food and Drug Administration (FDA) Center for Drug Evaluation and Research and Center for Biologics Evaluation and Research, become new therapeutic drugs (NTDs), each NTD can have an average annual sale worth $1.2 billion (Schulze, Ringel, Panier, & Baedeker, 2017). It has also been estimated that the cost of producing a new drug for sale is $2588 million (DiMasi, Grabowski, & Hansen, 2016). In this market the FDA considers pharmaceutical innovation in the following categories (Suri & Banerji, 2016):

1. New chemical entities.
2. New salts of previously approved drugs.
3. New formulations of previously approved medications.
4. New combinations of two or more medications.
5. Duplication by new manufacturers of products already marketed.
6. New indication for a medicine already marketed.
7. Medications not previously approved.

The largest global pharmaceutical companies in terms of the development of pharmaceutical innovations are GlaxoSmithKline, Pfizer, and Bristol Mayer Squibb; these represent the largest number of New Approved Medications (see Table 3.2) (Sellés, 2014). Their percentage of sales does not correlate with their discoveries and investment in R&D and innovation (see Table 3.3) (Schuhmacher, Gassmann, & Hinder, 2016). According to Schuhmacher et al. (2016) this is mainly because of:

Table 3.2 Drugs under development by the pharmaceutical industry, 2011

	Preclinical	*Clinical*			*NAM*[a]	*Total*
		Phase I	*Phase II*	*Phase III*		
GlaxoSmithKline	0	45	34	16	10	112
Pfizer	27	7	28	20	10	92
Bristol Myers Squibb	17	14	13	8	6	58
Roche	20	7	14	11	3	55
Astra Zeneca	13	7	12	11	9	52
Novartis	0	11	12	13	7	43
Bayer	19	9	7	3	2	40
Merck	13	3	2	14	2	34
Abbot	4	8	11	6	4	33
Johnson & Johnson	0	0	5	11	0	16

[a]New approved medications

Table 3.3 Top 10 pharmaceutical companies by sales (billions of dollars)

Pharmaceutical company	*2016*	*2015*
Johnson & Johnson	$71.89 billion	$70.04 billion
Pfizer	$52.82 billion	$48.85 billion
Roche	$50.11 billion	$47.70 billion
Novartis	$39.8 billion	$39.5 billion
Merck & Co.	$36.57 billion	$36.73 billion
Sanofi	$36.57 billion	$36.73 billion
GlaxoSmithKline	$34.79 billion	$29.84 billion
Gilead	$30.39 billion	$29.15 billion
AbbVie	$25.56 billion	$22.82 billion
Bayer	$25.27 billion	$24.09 billion

1. Lack of reproducibility of results published by researchers in terms of basic science.
2. Missing adequate pharmacokinetic studies.
3. Preclinical models poorly predictive for the discovery of new compounds.
4. Increase in complexity in the model of therapeutic targets for the discovery of new drugs.
5. Increase in the complexity of clinical studies.

Therefore, an initiative concerning bioprospecting for the country that will have a broad impact requires the implementation and standardization of preclinical tests that have a high predictive value, reproducibility, specificity, and precision (ICH Harmonized Tripartite, 2005). The purpose of these trials is to evaluate and give an added value to biodiversity with a criterion of truthfulness, in order to detect the greatest number of NEQs with which is possible to develop promising NTDs.

In this way, the growth of imports within the world market of medicines during the period 2001–2015 has gone from $38 billion to $186.4 billion, and in the same period has experienced a growth from $34.9 trillion to $186.8 trillion in terms of exports (Map, 2015). Moreover, the world trade in medicines (including veterinary drugs) in the period 2001–2015 has grown in imports from $96.4 trillion to $363.9 trillion, with exports increasing from $97.9 trillion to $336.4 trillion, with European countries developing the largest producers (Map, 2015). Medicines are one of the largest global markets, and increasing participation in innovation is necessary thanks to the evaluation, discovery, and development of new drugs. Likewise, the world market in pesticides and disinfectants has had a remarkable growth, represented by new bioactive molecules and products.

The FDA has approved 293 NCEs of the 358 that were submitted in the period 2006 to 2015 (FDA, 2016). But in 2016 the approval rate was the lowest for five years dropping to 23 NCEs, which, as well as being worrying, has also been a boost to investment in R&D and investment in novel sources of NCEs (Griesenauer & Kinch, 2017). On average, the pharmaceutical industry creates around 30 products per year, of which 25 correspond to NCEs and 5 are biological (Torre & Albericio, 2017). In addition, the sale of differentiated generics constituted a market worth $15 trillion in 2013, with a growth recorded in 2015 of $22 trillion to a total of $37, which opens possibilities for the development of supergenerics. These are defined as those new formulations that increase the dose of available drugs as well as their spectrum of activity, thereby decreasing the number of drug interactions. This is also considered pharmaceutical innovation according to the FDA (Suri & Banerji, 2016).

Alongside these advances, the pharmaceutical industry is reducing the risk factor in the development of new medicines by using a model in which the research units are grouped in global innovation bioclusters. These are located in San Francisco, Boston, London, Basel, and Shanghai (Gautam & Pan, 2016), places where the commercialization of new compounds coming from biodiversity is carried out. These are all potential clients for NCEs from Colombia's mega-diversity, but for this to be taken forward a sustainable and rational bioprospecting initiative that has real impact is necessary.

The sale of R&D services is a thriving source of funding for research centers, having gone from being a market of $14 trillion in 2003 to $47 trillion in 2011 (Billette de Villemeur & Versaevel, 2017). This represents a great business opportunity for studies in biodiversity and the evaluation of new products in conjunction with industries, owing to the need to reduce costs and increase the return margins when bringing chemically novel compounds to the market: these returns have increased from $451 million in 1991 to $2558 million in 2016 (Billette de Villemeur & Versaevel, 2017).

From another perspective, the market for phytotherapeutic products is a further source of economic growth that the country can access, because Colombia belongs to the International Convention for the Protection of New Varieties of Plants. This protects the use and cultivation of different varieties of plants for the benefit of the communities, likewise the European Union does not authorize patents for varieties of plants and animals (Sahoo & Bhattacharya, 2017). Currently the greatest trends in innovation in this field are directed towards the development of nutraceutical and cosmetic products (see Fig. 3.2) (Superintendencia de Industria y Comercio-SIC, CIGEPI, 2015).

In Colombia, the countries with the most patent applications registered on phytotherapeutics are the United States, Germany, and Malaysia. Colgate Palmolive is the largest with 14 applications. It is a priority to strengthen the national pharmaceutical industry in order to improve these low rates (see Figs. 3.3, 3.4, 3.5) (Superintendencia de Industria y Comercio-SIC, CIGEPI, 2015).

Johnson & Johnson Ltd. and Procter & Gamble have the equal largest number of phytotherapeutic registered brands in the national market, a fact that shows the business potential of this type of industry for the economic development of the country (see Fig. 3.6) (Superintendencia de Industria y Comercio-SIC, CIGEPI, 2015).

The development of new drugs is not an activity that is just focused on the discovery of novel molecules, but it also relates to the investigation of new applications and formulations. An effective bioprospecting model must take into account this broad front of activity.

Another promising market for Colombian biodiversity is that of fertilizers and pesticides for agricultural use. The current biological market (including plant extracts) is worth approximately $2.9 trillion. In an agro-

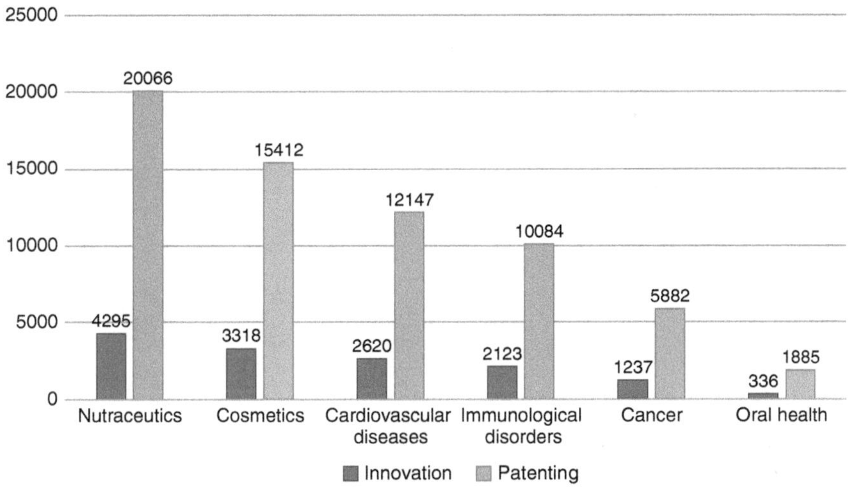

Fig. 3.2 Innovation trends in phytotherapeutics (number of products developed) (Superintendencia de Industria y Comercio-SIC, CIGEPI, 2015)

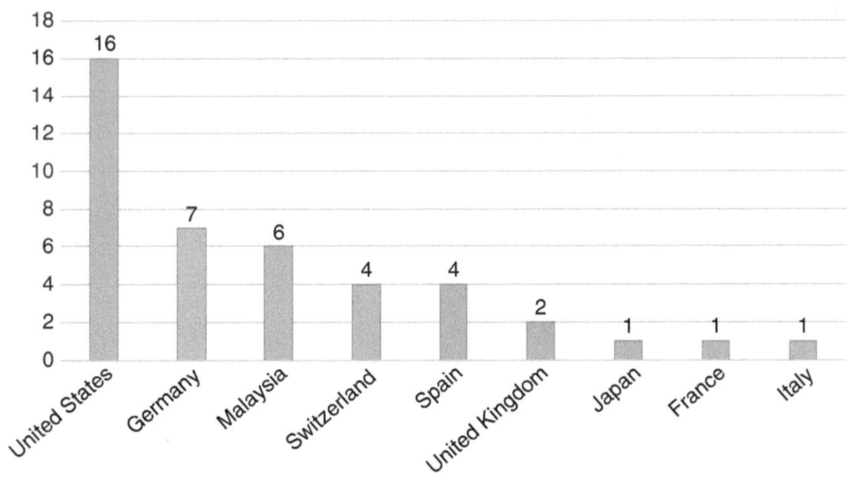

Fig. 3.3 Phytotherapeutic patent applicants in Colombia (Superintendencia de Industria y Comercio-SIC, CIGEPI, 2015)

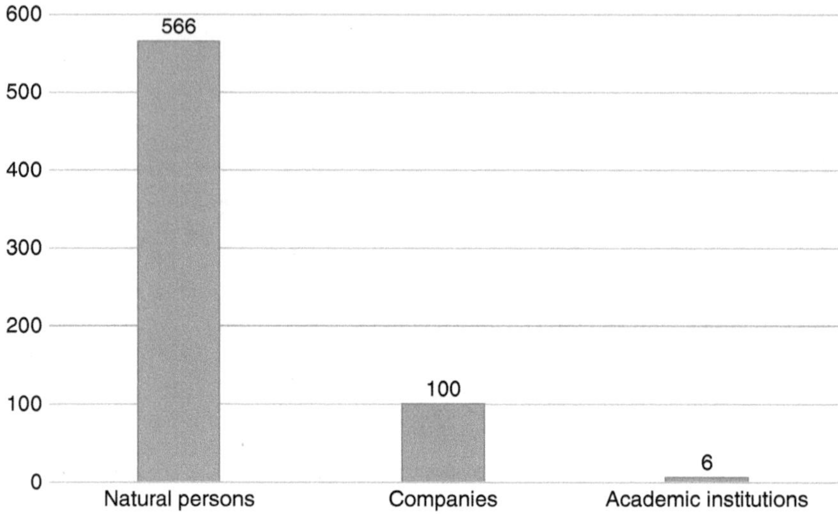

Fig. 3.4 Types of patent applicants for phytotherapeutics in Colombia (SIC, CIGEPI, 2015)

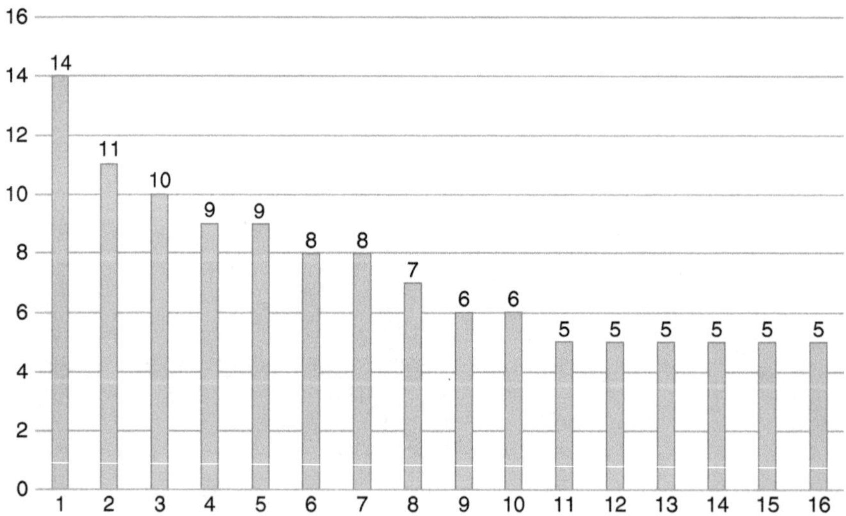

Fig. 3.5 Companies applying for patents on phytotherapeutics in Colombia (Superintendencia de Industria y Comercio-SIC, CIGEPI, 2015)

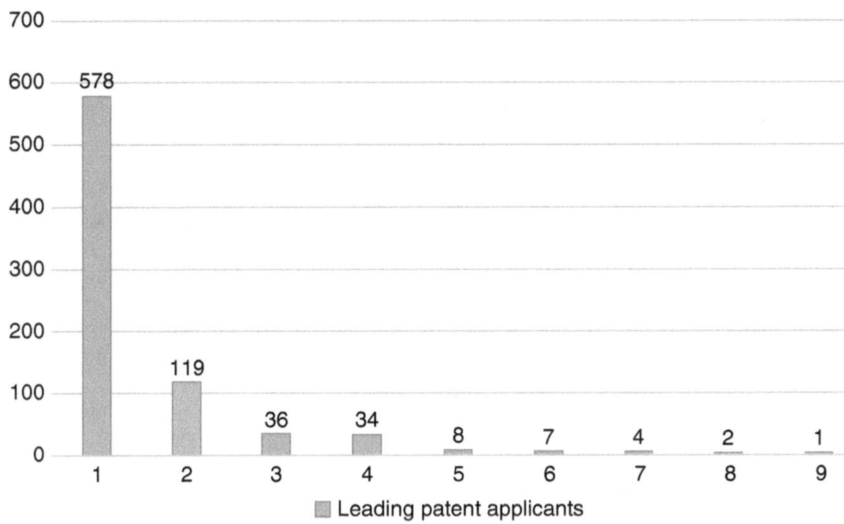

Fig. 3.6 Trademarks registered in Colombia by leading patent applicants on phytotherapeutic (Superintendencia de Industria y Comercio-SIC, CIGEPI, 2015)

chemical market valued at about $240 trillion, biopesticides represent $2 trillion per year, with 50% corresponding to products derived from microorganisms (Singh, 2017).

It is important to emphasize here that in Colombia the development of innovative medicines is only just beginning to happen, in spite of the tax benefits granted and an exemption from income tax on new medicinal products made in the country, despite this being a market that grew 7% in 2016 with annual sales of 3 billion Colombian pesos (Hernández & Miranda, 2015; Mendoza-Ruiz, Acosta, Escamilla, & Torres, 2017). The structure of the institutional market is focused on the production of generics, which means the innovation process is considered to be complex and expensive (Hernández & Miranda, 2015; Mendoza-Ruiz et al., 2017). This market is mostly represented by the sale of anti-infection drugs, according to figures relating to 1993–2007 (see Table 3.4) (Vásquez Velásquez, Gómez Portilla, & Rodríguez Acosta, 2010).

The increase in exports has made this market one of the most profitable in the country, with a growth of 17.4% in 2015 (Villar, Salazar, Pérez, Orbegozo, & Mesa, 2015). Likewise, the six largest pharmaceutical cor-

Table 3.4 Concentration of drug market in Colombia, 1993–2007 (Vásquez Velásquez et al., 2010)

Therapeutics groups	*Number of active substances*	*Number of pharmaceutical presentations*	*Number of laboratories*	*Market concentration*	*USD prices*
Antigouts	1	1	2	High	0.52
Antihistamines	2	3	9	High and medium	0.74
Anti-infectives	28	49	11	High and medium	2.07
Anti-inflammatories	4	7	8	High and medium	0.78
Cardiovascular and renal agents	8	14	12	High, medium and low	0.63
Gastrointestinal agents	6	7	9	High and medium	0.9
Hormones and hormonal regulators	3	3	3	High and medium	0.77
Skin and mucous membranes	3	6	8	High and medium	0.23
Respiratory diseases	5	8	8	High and medium	0.37
Nervous system agents	5	7	9	High and medium	0.49

porations in the country, according to their level of sales, represented 38% of the total market in 2015. For the 12 months to May 2015, the highest growth in sales was recorded by Abbott and Lafrancol, with an annual increase of 17.7% and 14.4% respectively (Villar et al., 2015).

Likewise, it is worth highlighting concerns regarding the increasing trend in the Colombian pharmaceutical industry to import finished products, in bulk or as semi-finished products. Therefore, an increase in the innovative profile of production is required (Díaz, Gallo, & Plazas, 2016).

3.6 Risk Analysis and Risk Management in a Bioprospecting Program

It is also necessary to evaluate the possible risks for bioprospecting programs related to the search for biotechnological applications. In the world the major failures are due to the lack of an approach that allows for the articulation of a sociotechnological network that manages and develops research aimed at innovation. Likewise, there is no adequate evaluation of the technical and scientific capacities necessary in order to carry out an adequate biodiversity screening (Amandine, 2010). Taking this into account, the risk management of a bioprospecting project should focus on three major areas:

1. Administrative and managerial risks.
2. Financial risks.
3. Scientific risks.

Likewise, the identification of possible risks is a priority before any project is executed (Kerzner & Kerzner, 2017). In this way it is possible to show that these risks can be broken down as follows:

1. Administrative and managerial risks:
   a. Quality management system, ISO 9001: 2015 standard.
   b. Implementation of ISO 17025 and 15189 norms.
   c. Project management.
   d. Measurement of the regional impact of the project.
2. Financial risks:
   a. Financing and cash flow.
   b. Products and services to be developed within the framework of the project.
   c. Sustainability.
   d. Creation of spin-off companies.
   e. Marketing and patents.
3. Scientific risks:
   a. Collection and selection of biological material (researchers insist on evaluating material that is not chemically diverse).
   b. Processing and taxonomic classification of biological material.
   c. Extraction of biological material for evaluation.
   d. Evaluation of biological activity, implementation of protocol CLSI- EP15-A2—User Verification of Performance for Precision and Trueness; Approved Guideline, Second Edition, to evaluate the reliability and precision of the tests in order to examine their verification and validation (Chesher, 2008).
   e. Evaluation of the toxicity of the identified material.
   f. Chemical analysis of active biological material.
   g. Productive staggering of biotechnological processes.

Correct evaluation and risk management will achieve an increase in the quality of the investigations, decreasing the time and cost of projects and increasing their success (Cagliano, Grimaldi, & Rafele, 2015).

3.7 Trends in Bioprospecting to Solve Problems in Innovation

One of the great problems in bioprospecting programs is achieving productive staggering of the secondary metabolites, as well as obtaining productive organisms and ensuring their growth, in order to make processes more cost effective (Firn, 2003). This problem can be solved through the use of synthetic biological methods, which allows the use of genes: these encode the production of the compounds that are of interest and allows them to be inserted into organisms that are more easily cultured; thus it is necessary to identify the material necessary for the application of these technologies (Unkles, Valiante, Mattern, & Brakhage, 2014). Developing these ideas, it may be noted that the detection of groups of biosynthetic genes, isolated from different genomes, will be very useful for new drug discovery strategies that are independent of the isolated producing microorganism (Hover, Kim, Katz, Charlop-Powers, et al., 2018). This is how natural products such as artemisinin can be produced by metabolic engineering for the marketing of pharmaceutical agents (Paddon & Keasling, 2014), giving greater added value to both the country's agencies and to these agencies' genetic resources.

Bionanotechnology is another great tool for the development of bioprocesses to enhance biodiversity, through the detection and evaluation of new organisms that are capable of producing nanoparticles and nanomaterials useful in medicine, agriculture, and industry (Bueno, 2017b). Likewise, these nanoparticles can enhance the development of new and varied bioactive products when used to obtain new formulations (Bueno, Demirci, & Baser, 2017).

3.8 Conclusions

The spin-off model for a nascent Colombian biopharmaceutical industry must be viewed from two perspectives: the university and the biopharmaceutical firm. The findings suggest the difference between these two business models: while the first emphasizes the scientific value of its knowledge through publications, the second emphasizes the commercial value of knowledge through patents (Erden, 2017).

The biopharmaceutical industry may be considered to be a tripartite vertical alliance network, where biotechnology firms license knowledge to universities for further development and technology transfer to large companies, as these have the resources for clinical studies and commercialization (Balconi & Lorenzi, 2017).

Equally, it is necessary to take into account the fact that any initiative in bioprospecting and biotechnology must subscribe to the corresponding contract for access to genetic resources and their derivative products. The Convention on Biological Diversity (CDB), adopted by Law 165 of 1994, recognizes the sovereign rights of states over their natural resources in areas within their jurisdiction. Specifically, the third objective of the CBD (developed in Articles 8j, 15, and 16) establishes fair and equitable participation in the benefits derived from the use of genetic resources, through, among other things, adequate access to those resources and an appropriate transfer of relevant technologies, taking into account all rights to these resources and technologies, as well as appropriate funding. In July 1996, the five Andean countries (Bolivia, Colombia, Ecuador, Peru, and Venezuela) signed the Andean Decision 391 as part of the Board of Cartagena Agreement.

The purpose of the Andean Decision is to regulate access to genetic resources and their by-products or their intangible components in order to:

1. Guarantee a fair and equitable participation of the benefits derived from the use of resources and traditional knowledge associated with them.
2. Set the bases for the recognition and valuation of genetic resources and their derived products.
3. Promote the conservation of diversity and the development of scientific, technological, and technical capabilities.
4. Strengthen the negotiating capacity of member countries.

Finally, under the concept of total quality management, the initiative for implementing bioprospecting and biotechnology centers for biodiversity study should be undertaken by an organization that takes into account the following structure (Feigembaun, 1991).

Strategies: Define policies, objectives, and guidelines for the achievement of quality and customer satisfaction. These policies and objectives must be aligned with the results that the organization wishes to obtain.

Processes: The processes, activities, and procedures required for the realization of the product or service must be determined, analyzed, and implemented, and in turn aligned with the objectives set. The monitoring and control activities for the efficient operation of the processes must also be defined.

Resources: Define clear assignments of personnel, equipment, and/or machinery necessary for the production or provision of the service, the work environment, and the financial resource necessary to support quality activities.

Organizational Structure: Define and establish a structure of responsibilities, authorities, and flow of communication within the organization.

Documents: Establish the procedures, documents, forms, registers, and any other documentation for the effective and efficient operation of the processes, and therefore of the organization. In this order of ideas, the bioprospecting and biotechnology laboratories for biodiversity study should be certified under the following standards: ISO/IEC 17025, ISO 15189, and CLSI EP15-A2 for the accreditation of biological analysis.

References

Arundel, A., & Sawaya, D. (2009). *The Bioeconomy to 2030: Designing a Policy Agenda*. Retrieved from http://www.oecd.org/futures/longtermtechnologicalsocietalchallenges/thebioeconomyto2030designingapolicyagenda.htm

Amandine, B. (2010). Technological Choices in International Environmental Negotiations: An Actor—Network Analysis. *Business & Society, 49*(4), 570–590.

Balconi, M., & Lorenzi, V. (2017). The Increasing Role of Contract Research Organizations in the Evolution of the Biopharmaceutical Industry. *African Journal of Business Management, 11*(18), 478–490.

Billette de Villemeur, E., & Versaevel, B. (2017). *One Lab, Two Firms, Many Possibilities: On R&D outsourcing in the Biopharmaceutical Industry*. Retrieved from https://mpra.ub.uni-muenchen.de/76903/1/MPRA_paper_76903.pdf

Bueno, J. (2017a). Bioprospecting and Drug Development, Parameters for a Rational Search and Validation of Biodiversity. *Journal of Microbial and Biochemical Technology, 9*, e128.

Bueno, J. (2017b). Fungal Bionanotechnology, When Knowledge Merge into a New Discipline to Combat Antimicrobial Resistance. In *Fungal Nanotechnology* (pp. 189–206). Cham: Springer.

Bueno, J., Demirci, F., & Baser, K. H. C. (2017). Antimicrobial Strategies in Novel Drug Delivery Systems: Applications in the Treatment of Skin and Soft Tissue Infections. In *The Microbiology of Skin, Soft Tissue, Bone and Joint Infections* (pp. 271–286). Boston: Academic Press.

Cagliano, A. C., Grimaldi, S., & Rafele, C. (2015). Choosing Project Risk Management Techniques. A Theoretical Framework. *Journal of Risk Research, 18*(2), 232–248.

Cheng, X., Zhao, S. G., Xiao, X., & Chou, K. C. (2017). iATC-mHyb: A Hybrid Multi-label Classifier for Predicting the Classification of Anatomical Therapeutic Chemicals. *Oncotarget, 8*(35), 58494.

Chesher, D. (2008). Evaluating Assay Precision. *The Clinical Biochemist Reviews, 29*(Suppl 1), S23.

Colombia, C. N. Ley 1753 de 2015 Por la cual se expide el Plan Nacional de Desarrollo 2014–2018 todos por un nuevo país. Bogotá DC: DNP Departamento Nacional de Planeación. Retrieved from https://colaboracion.dnp.gov.co/CDT/PND/PND%202014-2018%20Tomo%201%20internet.pdf

Consejo privado de competitividad. (2016). Índice departamental de competitividad. Retrieved from https://compite.com.co/wp-content/uploads/2016/07/CPC_IDC-2016.pdf

Cotes Prado, A. M., Meneses, B., Villamizar, L. S. R., Mogollón, F. Z., Martínez, M. V. A., & Prado, H. A. M. C. (2012). Bioprospección para el desarrollo del sector agropecuario de Colombia (No. Doc. 26007) CO-BAC, Bogotá.

De Witt, A., Osseweijer, P., & Pierce, R. (2017). Understanding Public Perceptions of Biotechnology Through the "Integrative Worldview Framework". *Public Understanding of Science, 26*(1), 70–88.

Departamento Administrativo de Ciencia, Tecnología e Innovación (COLCIENCIAS). (2017). Resolución 0036 de 2017 por "La cual se adoptan los lineamientos de la presentación de proyectos de oferta institucional de inversión en Ciencia, Tecnología e Innovación". Retrieved from http://www.colciencias.gov.co/sites/default/files/upload/reglamentacion/resolucion-0036-2017.pdf

Departamento Nacional de Planeación. (2011). Documento CONPES 3697, Política para el desarrollo comercial de la biotecnología a partir del uso sostenible de la biodiversidad. Retrieved from https://www.cbd.int/doc/measures/abs/post-protocol/msr-abs-co-es.pdf

DiMasi, J. A., Grabowski, H. G., & Hansen, R. W. (2016). Innovation in the Pharmaceutical Industry: New Estimates of R&D Costs. *Journal of Health Economics, 47*, 20–33.

Domínguez Molina, Á. M. (2012). *Aproximación al estado actual de la bioprospección en Colombia y propuesta para la implementación del plan nacional en bioprospección*. Master's thesis, Facultad de Estudios Ambientales y Rurales. Retrieved from https://repository.javeriana.edu.co/bitstream/handle/10554/12379/DominguezMolinaAngelaMaria2012.pdf?sequence=1

Díaz, B. M. V., Gallo, J. J., & Plazas, C. E. (2016). Propuesta de una metodología para el estudio de competencia tecnológica, en empresas fabricantes de medicamentos. *Universidad & Empresa, 18*(31), 11–27.

Erden, Z. (2017). The Publishing and Patenting Strategies of Successful University Spinoffs in the Biopharmaceutical Industry. *Drug Discovery Today, 22*(1), 5–9.

Feigembaun, A. V. (1991). Key Systems Activities for Total Quality Control. In A. V. Feigembaun (Ed.), *Total Quality Control* (p. 94). Estados Unidos: McGraw-Hill.

Firn, R. D. (2003). Bioprospecting–Why Is It So Unrewarding? *Biodiversity & Conservation, 12*(2), 207–216.

Food and Drug Administration. (2016). *Novel New Drugs, 2015 Summary.* Retrieved from https://www.fda.gov/downloads/drugs/developmentapprovalprocess/druginnovation/ucm481709.pdf

Gallón, A. I. M. (2011). La biodiversidad y la biotecnología en la facultad de química farmacéutica de la Universidad de Antioquia, a propósito del documento conpes 3697 de 2011. *Vitae, 18*(2), 111–113.

Gautam, A., & Pan, X. (2016). The Changing Model of Big Pharma: Impact of Key Trends. *Drug Discovery Today, 21*(3), 379–384.

Gonzales, C., Villa, J. M., & Velasco, R. (2007). Biotecnología: desde el punto de vista de los negocios. *Facultad de Ciencias Agropecuarias, 5*(1), 36–41.

Goyes, D. R., & South, N. (2015). Land-grabs, Biopiracy and the Inversion of Justice in Colombia. *British Journal of Criminology, 56*(3), 558–577.

Griesenauer, R. H., & Kinch, M. S. (2017). 2016 in Review: FDA Approvals of New Molecular Entities. *Drug Discovery Today, 22*(11), 1593–1597.

Hernández, M. C. (2008). Propuesta de apoyo para una Gestión eficiente de la biotecnología. *Revista Escuela de Administración de Negocios, 62.*

Hernández, R. S., & Miranda, P. P. (2015). Una mirada a la industria farmacéutica en Colombia. *REVISTA FACCEA, 4*(2), 107–115.

Hover, B. M., Kim, S. H., Katz, M., Charlop-Powers, Z., et al. (2018). Culture-independent Discovery of the Malacidins as Calcium-dependent Antibiotics with Activity Against Multidrug-resistant Gram-positive Pathogens. *Nature Microbiology, 3*(4), 415–422.

ICH Harmonized Tripartite. (2005). *Validation of Analytical Procedures: Text and Methodology. Q2 (R1)*, Vol. 1.

Jiménez, T., Cárdenas, J., & Soler-Tovar, D. (2017). Biocomercio en el contexto suramericano: Colombia y Perú como estudios de caso. *Revista de Medicina Veterinaria, 35*, 9–15.

Kang, K. N., Jung, C. S., & Ryu, T. K. (2015). The Nagoya Protocol and the Biotechnology Industry. *International Journal of Pharma Medicine and Biological Sciences, 4*(3), 209.

Kamuriwo, D. S., & Baden-Fuller, C. (2016). Knowledge Integration Using Product R&D Outsourcing in Biotechnology. *Research Policy, 45*(5), 1031–1045.

Kerzner, H., & Kerzner, H. R. (2017). *Project Management: A Systems Approach to Planning, Scheduling, and Controlling.* New York: John Wiley & Sons.

Kusari, S., Singh, S., & Jayabaskaran, C. (2014). Biotechnological Potential of Plant-associated Endophytic Fungi: Hope Versus Hype. *Trends in Biotechnology, 32*(6), 297–303.

Map, T. (2015). *Trade Statistics for International Business Development.* In Geneva, Switzerland: International Trade Centre (United Nations Conference on Trade and Development-World Trade Organization). Retrieved from http://www.trademap.org

Marín, C., & Parra, S. (2015). *Bitácora de flora. Guía visual de plantas de páramos de Colombia.* Retrieved from repository.humboldt.org.co/bitstream/handle/20.500.11761/9283/BitacoraFLORA-Agosto11-Final_Digital.pdf?sequence=1&isAllowed=y

Mazzucato, M. (2015). The Green Entrepreneurial State. In I. Scoones, M. Leach, & P. Newell, (Eds.), *The Politics of Green Transformations* (pp. 134–152). London: Routledge.

Mendoza, C. A. S. (2014). Análisis de los indicadores de biocomercio en America (Colombia). *Asuntos Económicos y Administrativos, 27*, 331–341.

Mendoza-Ruiz, A., Acosta, A., Escamilla, E. P. V., & Torres, M. C. L. (2017). Pharmaceutical Policy in Colombia. In *Pharmaceutical Policy in Countries with Developing Healthcare Systems* (pp. 193–219). Springer International Publishing.

Ministerio de Agricultura y Desarrollo Rural. (2017). Plan Estratégico de Ciencia Tecnología e Innovación del Sector Agropecuario Colombiano para la Agroindustria (PECTIA) y Sistema Nacional de Innovación Agropecuario (SNIA). Retrieved from http://www.colombiacompetitiva.gov.co/sncei/Documents/pectia-terminado.pdf

Montero, J. M. S. (2011, December). Biotecnología: presente y futuro. In *Anales de la Real Academia Nacional de Farmacia* (Vol. 77, No. 4).

Moreno, L. A., Andrade, G. I., & Ruíz-Contreras, L. F. (Eds.). (2016). *Biodiversidad 2016. Estado y tendencias de la biodiversidad continental de Colombia.* Instituto de Investigación de Recursos Biológicos Alexander von Humboldt. Bogotá, DC, Colombia, p. 106. Retrieved from http://repository.humboldt.org.co//handle/20.500.11761/32962

Moscoso, F., Estrada, F., Diaz, N., & Andrade, N. (2015). Obstacles to Progress in R&D Activities Caused by Institutional and Regulatory Frameworks: The Case of the Biotech Sector in Colombia. *Research in World Economy, 6*(4), 116.

Paddon, C. J., & Keasling, J. D. (2014). Semi-synthetic Artemisinin: A Model for the Use of Synthetic Biology in Pharmaceutical Development. *Nature Reviews Microbiology, 12*(5), 355.

Pan, P. G. (2006). *Bioprospecting: Issues and Policy Considerations.* Legislative Reference Bureau. Retrieved from http://www.angelfire.com/planet/big60/BioprospectingCommitsionReport.pdf

Prip, C., & Rosendal, K. (2015). Access to Genetic Resources and Benefit-sharing from Their Use (ABS)-state. *World, 9*(2), 189–212.

Purkayastha, J. (2016). Emerging Trends in Sustainable Bioprospecting of Bioresources. In *Bioprospecting of Indigenous Bioresources of North-East India* (pp. 3–19). Singapore: Springer.

Pushpangadan, P., Ijinu, T. P., Dan, V. M., & George, V. (2015). Trends in Bioprospecting of Biodiversity in New Drug Design. *Pleione, 9*(2), 267–282.

Pyka, A., & Prettner, K. (2018). Economic Growth, Development, and Innovation: The Transformation Towards a Knowledge-Based Bioeconomy. In *Bioeconomy* (pp. 329–340). Cham: Springer.

Quezada, F. (2003). *Biotechnology-based Opportunities for the Sustainable Use of Biodiversity Resources in the Andean Region: Recommendations and Strategic Guidelines.* Retrieved from http://citeseerx.ist.psu.edu/viewdoc/download;jsessionid=153B57C474F06F5FC1C3DC9B7B0536B3?doi=10.1.1.488.170&rep=rep1&type=pdf

Rangel, J. O. (2015). La biodiversidad de Colombia: significado y distribución regional. Palimpsestvs, Revista de la Academia Colombiana de Ciencias Exactas. *Físicas y Naturales, 39*(151), 176–200.

Rodríguez, J. M., Camargo, J. C., Niño, J., Pineda, A. M., Arias, L. M., Echeverry, M. A., & Miranda, C. L. (2009). *Valoración de la Biodiversidad en la Ecorregión del Eje Cafetero.* Pereira: CIEBREG.

Sahoo, N., & Bhattacharya, S. S. (2017). Intellectual Property Rights Protection in Plants: Scopes in Lychee Commercialization. In *The Lychee Biotechnology* (pp. 281–299). Singapore: Springer.

Schaffartzik, A., Haberl, H., Kastner, T., Wiedenhofer, D., Eisenmenger, N., & Erb, K. H. (2015). Trading Land: A Review of Approaches to Accounting for Upstream Land Requirements of Traded Products. *Journal of Industrial Ecology, 19*(5), 703–714.

Schuhmacher, A., Gassmann, O., & Hinder, M. (2016). Changing R&D Models in Research-based Pharmaceutical Companies. *Journal of Translational Medicine, 14*(1), 105.

Schulze, U., Ringel, M., Panier, V., & Baedeker, M. (2017). Market Watch: Value of 2016 FDA Drug Approvals: Reversion to the Mean? *Nature Reviews Drug Discovery, 16*(2), 78–78.

Sellés, A. J. N. (2014). Pharma R&D in Latin America. *Pharmaceuticals Policy and Law, 16*(3, 4), 267–275.

Jr Silva, G., Silveira, J. M., Paranhos, J., Hasenclever, L., & Miranda, R. (2017). Innovation, Market Power and Biotechnology in the Brazilian Chemical Industry. *Economics Bulletin, 37*(2), 1210–1216.

Singh, B. K. (2017). Creating New Business, Economic Growth and Regional Prosperity Through Microbiome-based Products in the Agriculture Industry. *Microbial Biotechnology, 10*(2), 224–227.

Srivastava, S. K. (2016). *Commercial Use of Biodiversity: Resolving the Access and Benefit Sharing Issues*. New Delhi: SAGE Publications India.

Superintendencia de Industria y Comercio-SIC, Centro de Información Tecnológica y Apoyo a la Gestión de la Propiedad Industrial – CIGEPI. (2015). *Productos fitoterapéuticos*. Retrieved from http://www.sic.gov.co/recursos_user/documentos/publicaciones/Productos_Fitoterapeuticos.pdf

Suri, F. K., & Banerji, A. (2016). Super Generics—First Step of Indian Pharmaceutical Industry in the Innovative Space in US Market. *Journal of Health Management, 18*(1), 161–171.

Torre, B. G., & Albericio, F. (2017). The Pharmaceutical Industry in 2016. An Analysis of FDA Drug Approvals from a Perspective of the Molecule Type. *Molecules, 22*(3), 368.

Unkles, S. E., Valiante, V., Mattern, D. J., & Brakhage, A. A. (2014). Synthetic Biology Tools for Bioprospecting of Natural Products in Eukaryotes. *Chemistry & Biology, 21*(4), 502–508.

Vargas-Hernández, J. G., Pallagst, K., & Hammer, P. (2017). Bio Economy's Institutional and Policy Framework for the Sustainable Development of Nature's Ecosystems. *Atlantic Review of Economics, 2*, 1–30.

Villar, L., Salazar, N., Pérez, C., Orbegozo, C., & Mesa, C. A. (2015). *Informe del sector farmacéutico colombiano-FEDESARROLLO-ANDI*. Retrieved from http://www.repository.fedesarrollo.org.co/bitstream/handle/11445/2481/Repor_Julio_2015_Fedesarrollo_y_ANDI.pdf?sequence=3&isAllowed=y

Vásquez Velásquez, J., Gómez Portilla, K., & Rodríguez Acosta, S. (2010). Regulación en el mercado farmacéutico colombiano. *Revista de Ciencias Sociales (Ve), 16*(2).

Wyatt, T., & Brisman, A. (2017). The Role of Denial in the 'Theft of Nature': Comparing Biopiracy and Climate Change. *Critical Criminology, 25*(3), 325–341.

Zanello, G., Fu, X., Mohnen, P., & Ventresca, M. (2016). The Creation and Diffusion of Innovation in Developing Countries: A Systematic Literature Review. *Journal of Economic Surveys, 30*(5), 884–912.

CHAPTER 4

Patents for All: A Content Analysis of an Open-access Dataset of Colombian Patents 1930–2000

Julián David Cortés-Sánchez

4.1 Introduction

The sustainable wealth of nations is now intangible. While ***produced capital*** (i.e., machinery, structures, and equipment) and ***natural capital*** (i.e., agricultural land, forest, and minerals) both combined represent an estimate of 20–40% of wealth in most countries, ***intangible capital*** (i.e., human, social, and institutional capital) represents an estimate of 60–80% (the World Bank, 2011). Components such as the rule of law and intellectual property rights—both expressions of institutional capital—are crucial for sustainable economic growth (Hasan & Tucci, 2010; Hu & Png, 2010). Hence, both public and private sectors recognize the importance of investing in intangible capital for knowledge production stocks, such as the publication of patents (Hall, Jaffe, & Trajtenberg, 2005; Jaffe & Trajtenberg, 1996).

J. D. Cortés-Sánchez (✉)
School of Management, Universidad del Rosario, Bogotá, Colombia
e-mail: julian.cortess@urosario.edu.co

C. I. Pardo Martínez et al. (eds.), *Analysis of Science, Technology, and Innovation in Emerging Economies*,
https://doi.org/10.1007/978-3-030-13578-2_4

A patent is defined by the World Intellectual Property Organization (WIPO) as follows (WIPO, n.d.):

> … *[A]n exclusive right granted for an invention, which is a product or a process that provides, in general, a new way of doing something, or offers a new technical solution to a problem. To get a patent, technical information about the invention must be disclosed to the public in a patent application.*

To this day, there is no unified international system for patent granting. Patents' exclusive rights are only applicable in the country in which a patent was filed and granted. There are several regional patent offices which receive applications and grant patents for regional members (e.g., the African Intellectual Property Organization or the European Patent Office [EPO]). A path for multi-territorial patenting is via WIPO's Patent Cooperation Treaty. Any resident or national of any of the 152 member states of the Patent Cooperation Treaty (PCT) can file a single application which has the same legal enforcement as a national patent in some or all PCT member states (WIPO, n.d.).

The publication of patents worldwide has been growing steadily over the last 30 years. In 1985 the number of patents granted was 397,580; conversely in 2015 it was 1,241,000. More than triple. The average growth rate during this 30-year period was 4.3% (WIPO, 2015). However, there is considerable variation in national capabilities for patents publication. The number of applications by residents worldwide in 2015 was 1,862,548, from which 288,335 (15%) were from the United States and 321 (0.01%) from Colombia (WIPO, 2015).

Such disparities among developed and underdeveloped nations are partly explained by historical and institutional differences. For instance, the intellectual protection granted by patents during the Industrial Revolution was an inclusive institution (i.e., norm/rule [North, 1989]) that boosted the incentives for technological innovation and economic growth in England and Europe during the eighteenth and nineteenth centuries (Acemoglu & Robinson, 2013). In contrast, slavery was an extractive institution intensively used in Colombia for gold mining in the Pacific region during colonial times. Even today, those same regions are characterized by growing poverty and land tenure inequality, as well as lower school enrollment rates, vaccination coverage, and public goods provision (Acemoglu, García-Jimeno, & Robinson, 2012). Understanding the historical outcomes of institutions is crucial to adjust and improve institutions within public and private organizations, and public policies (Rennie, 2002). Furthermore, in a sense, historical standpoints can provide an overview of the outcomes supported or restricted by national and regional policies, and more than just for the sake of avoiding repetition, to liberate the future from its own tragic past (Harari, 2015).

In Colombia, the only public entity in charge of patent application and granting is the Colombian Supervisory Agency of Industry and Commerce (*Superintendencia de Industria y Comercio* [SIC]). Novelties such as products, procedures, production methods, machines, and devices produced by them can be patented. Conversely, discoveries such as new species or living beings existing in nature, scientific theories, mathematical models, therapeutic or surgical methods, financial or business plans, artworks (i.e., paintings, books) or software cannot be patented. The former novelties (i.e., products, procedures, production methods, and so forth) must fulfill three conditions to be considered for patenting: (1) it has to be new, anywhere in the world; (2) it should have inventiveness, beyond any obvious or irrelevant improvement; and (3) it should have a productive purpose or prone to be manufactured. Currently, the patent application procedure has eight stages (i.e., filing and admission, formal analysis, publication, third-party oppositions, patentability tests, in-depth test, final decision, and maintenance) and it may take three to five years to complete the process. Once finished, a utility model patent is granted for ten years, while an invention patent is granted for 20 years. If the inventor does not use the patent for three years, the patent expires and goes to the public domain. A patent also expires if the inventor does not pay the annual fee for maintenance.

The history-related literature on patents in Colombia, nevertheless, is negligible. One of the few peer-reviewed studies on the subject analyzed a sample of patents granted to at least one Colombian organization or individual from 1968 to 2007 (Sánchez, Medina, & León, 2007). No peer-reviewed studies, to our knowledge, investigate patents prior to 1968. Considering this, one of the most important and, paradoxically, barely known studies on patents and inventions in Colombia was published by Alberto Mayor back in 2005. Besides the historical and sociological analysis on patents and inventions, Mayor listed more than 1500 patents registered from 1930 to 2000 with their corresponding title, inventor (i.e., individual male/female inventor, group of inventors or companies), and place of origin (i.e., municipality or city).

The purpose of this chapter is to discuss the historical advancements and failures regarding patents in Colombia from 1930 to 2000, by identifying critical factors such as patent publication by year, the coefficient of invention, type of inventor (male, female, group, or company), geographical location, classification, and a content analysis of the text corpus of the patents' titles. By doing so, this chapter will provide an open-access dataset for further research or replication and public policy use, as often required by the scientific community (Open Science Collaboration, 2015). Besides replication, the scientific community is discussing triangulation as a multi-methodical and multi-disciplinary appraisal to address a particular research question. Results that agree across different methodologies or disciplines are less likely to be artifacts (Munafò & Smith, 2018). Considering that Mayor's analytical framework is mostly grounded in Science, Technology, and Society (STS) studies, the content analysis and text-mining applied to his patents database seek to identify and discuss underlying findings overlooked by solely using a qualitative approach.

After this introduction, we present a peer-reviewed literature review on Colombian patents. We then present the content analysis results based on Mayor's (2005) research. Then, we discuss the results with additional national and international studies. Finally, we present the conclusions of this study.

4.2 Literature Review

Peer-reviewed literature on Colombian patents is notably scarce, and diagnoses and results are neither pessimistic nor promising. The few studies found related to (1) intellectual property management models in Ibero-American universities, centered on patent publication (Del Socorro López Gómez et al., 2009); (2) networks of collaboration and patent publication by universities from the Community of Andean Nations (CAN) (Aguero, 2017); (3) the international publication of patents by Colombian inventors and organizations (Sánchez et al., 2007); (4) the impact of patent publication on economic growth in eight countries from Latin America (Robledo & Saavedra, 2016); and (5) immunology-related patent publication in Latin America (Campos & Campos, 2014).

Among the studies reviewed, Del Socorro López Gómez et al. (2009) noted an increasing awareness by Ibero-American universities (e.g., Universidad Nacional de Colombia and Universidad de Antioquia) regarding the need to strengthen their innovation and patents management model to contribute to knowledge-based economic growth. When comparing successful patent applications for 2005–2015 in CAN member universities (i.e., Bolivia, Colombia, Ecuador, Peru, and Venezuela), Colombian institutions stand out: from a total of 381 granted patents, 68.8% (262) were from Colombia, 28.8% (111) from Peru, 1.6% (6) from Ecuador, and 0.5% (2) from Bolivia (Aguero, 2017). From a historical perspective, Sánchez et al. (2007) pinpointed a sample of 223 patents endorsed by at least one organization/individual from Colombia from a sample of international patents for the period 1968–2007. Sánchez et al. (2007) also concluded that 59% of the patents were registered in the United States, 16% in Europe, and 25% in other countries. Furthermore, 3% (7) of the patents were granted to universities and 96.8% (216) to organizations in other sectors. The most prominent technological areas in patenting were identified as "electric digital data processing" (IPC [International Patent Classification]: G06F) and "filters implantable into blood vessels; prostheses; devices providing patency to, or preventing collapsing of, tubular structures of the body, e.g. stents; orthopedic, nursing or contraceptive devices; fomentation; treatment or protection of eyes or ears; bandages, dressings or absorbent pads; first-aid kits" (IPC: A61F). These findings showed fruitful milestones for Colombia in terms of patent publication and economic growth, considering that in the long term, a 10% increase in registered patents by residents could produce a 0.25% increase in GDP. The same increase in registered patents by non-residents could produce a 44% increase in GDP (Robledo & Saavedra, 2016).

Nevertheless, patenting processes within universities are not yet the main driver to strengthen the relationships between University–Industry–Government (Del Socorro López Gómez et al., 2009). At the regional level, Latin American countries represented 0.19% of the total patents registered in the United States Patent and Trademark Office (USPTO) in 2008–2012 (Aguero, 2017). These gaps also exist among Latin American countries: while Brazil sought the registration of 201 patents in 2006, Colombia applied for only 29 (Sánchez et al., 2007). Regardless of the fact that 96.8% of Colombian patents registered in 1968–2007 were granted to private and other organizations, none of Colombia's leading companies (i.e., Ecopetrol, Bavaria, Terpel, Grupo Nacional de Chocolates o Codensa) appeared in the top 20 (Campos & Campos, 2014). Furthermore, of the applications for immunology-related patents in 2004–2011, Colombia was ranked just 57th with 3 patent applications. In contrast, the United States applied for 10,464 patents during the same period (Campos & Campos, 2014).

In Colombia, patent registration has been regulated by the SIC since the 1970s and monitored by the Science and Technology Observatory

(OCyT) since the 1990s (SIC, n.d.-a; OCyT, n.d.). The technical reports published by the latter since 2004 contain information regarding Colombian patents from 1991 to 2016. They focus on the number of patent applications by residents and non-residents in national and international offices, the number granted, type of entity to which the patent was granted (e.g., private or public organization, natural person, etc.), city of registration, and economic sector (OCyT, 2004–2015). The most recent OCyT report (2015) outlined several remarkable and overall encouraging findings. Figure 4.1 presents the dependency rate for 2006–2015. The dependency rate represents the number of patent applications by non-residents divided by those by residents. The rate shows a continual downward trend, which means that the number of patents granted to residents has increased compared with non-residents.

Figure 4.2 presents the self-sufficiency rate for 2006–2015. The self-sufficiency rate represents the number of patent applications by residents divided by the total number of patent applications. This figure shows a continual upward trend, which means that the number of patent applications by residents has increased compared with the total number of patent applications.

Figure 4.3 presents the invention coefficient. The invention coefficient refers to the number of patent applications per 100,000 inhabitants. The figure shows a continual upward trend since 2011; this follows a period of stagnation between 2008 and 2010. Thus, the number of patent applications increased after normalizing it for the national population.

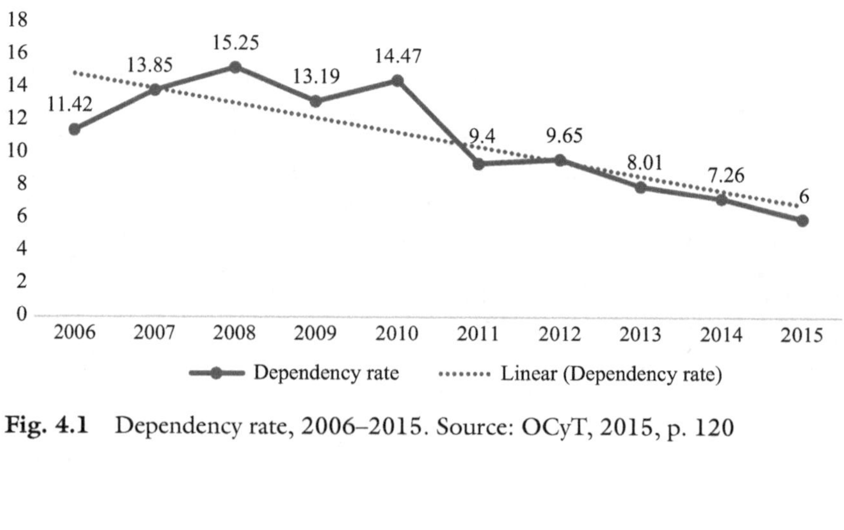

Fig. 4.1 Dependency rate, 2006–2015. Source: OCyT, 2015, p. 120

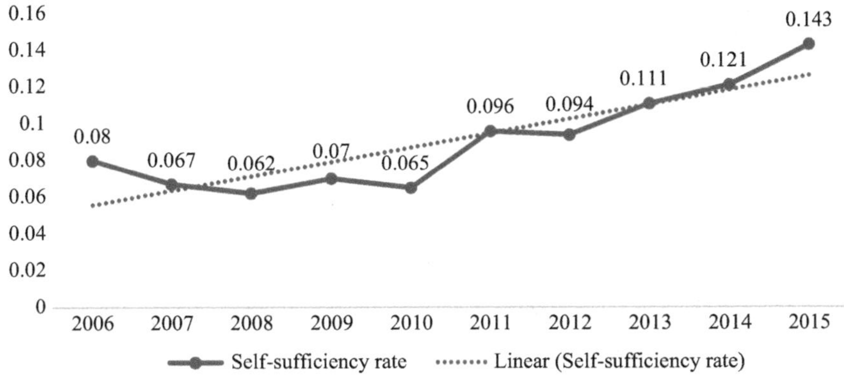

Fig. 4.2 Self-sufficiency rate, 2006–2015. Source: OCyT, 2015, p. 121

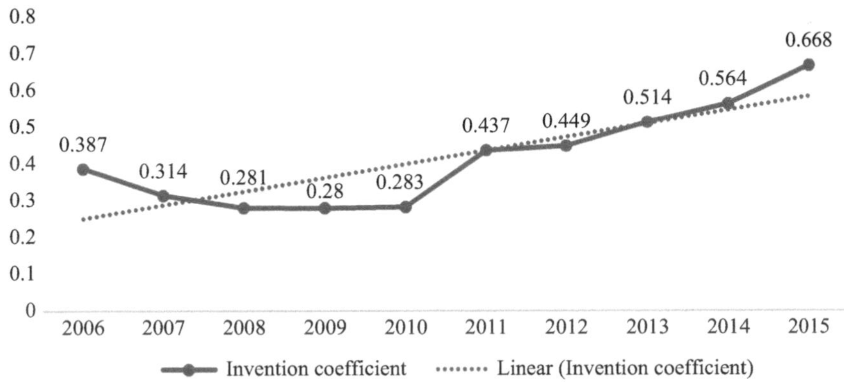

Fig. 4.3 Invention coefficient, 2006–2015. Source: Author, based on OCyT, 2015, p. 122

In Fig. 4.4 we can see Colombia's rate of patent applications over the number of patents granted in international offices. The following international offices were analyzed: USPTO, the EPO), and the Japan Patent Office (JPO). The performance of Colombian patent applications was mixed. The rate shows an upward trend in the USPTO. Furthermore, the 2015 rate (1.6) was more than triple that in 2014 (0.5). In contrast, both rates for EPO and JPO showed stagnation (EPO) and non-patents granted in several years (2006, 2008, and 2013) (JPO). It should be

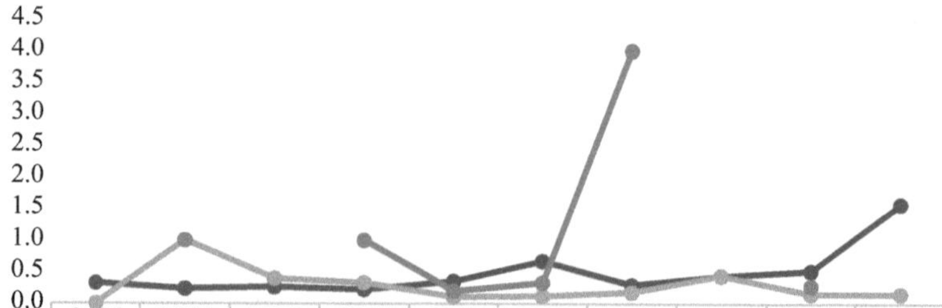

	2006	2007	2008	2009	2010	2011	2012	2013	2014	2015
USPTO	0.3	0.2	0.3	0.2	0.4	0.7	0.3	0.4	0.5	1.6
EPO	0.0	1.0	0.4	0.3	0.1	0.1	0.2	0.4	0.2	0.2
JPO		1.0		1.0	0.2	0.3	4.0		0.3	

USPTO EPO JPO

Fig. 4.4 Colombians' patent application and granted patents rates in international offices, 2006–2015. Source: Author, based on OCyT, 2015, p. 123

noted, however, that the 2012 rate for the JPO was very high (i.e., at 4.0 it was more than twice the highest rate in the USPTO).

As noted, Colombia has experienced an overall improvement at several levels regarding independence, self-sufficiency, invention, and patenting in international offices (USPTO, mainly). Nevertheless, these milestones are not enough to catch up to other regional countries (e.g., Brazil) or the global leaders (e.g., China, the United States, and South Korea).

The literature review on patents from a historical perspective in Colombia only identified one study, that by Sánchez et al. (2007), which focused on patent publication by (non-)residents from 1968 to 2007. Another look at the STS literature found one of the most important yet largely unknown and uncited studies from a historical perspective on patents and inventions in Colombia, namely, *Inventos y patentes en Colombia 1930–2000: de los límites de las herramientas a las fronteras del conocimiento* (Inventions and patents in Colombia 1930–2000: from the limits of tools to the knowledge frontiers) by Alberto Mayor (2005). In addition to the historical and sociological analysis on patents and inventions grounded in STS theory, Mayor listed more than 1500 patents registered between 1930 and 2000. Mayor's list contains information on type of inventor (male, female, group, or company), geographical location, and patent titles. He based his invention and patents list on four sources: (1) *Diario Oficial* (1930–1945) relatively complete statistical series; (2) *Diario Oficial* (1958–1957) mostly incomplete statistical series; (3) *Gaceta de la Propiedad Industrial* (1958–1976) complete statistical series; and (4) *Gaceta de la Propiedad Industrial* (1977–2000) incomplete statistical series. As noted, some of the sources were incomplete. Considering that the data and analysis on patents and inventions prior to 1968 was not found in peer-reviewed literature, the result of this archive workforce is thus outstanding. The content of this list is analyzed and discussed in the following sections.

4.3 Methods

Content analysis using Voyant Tools was implemented for Mayor's list. Voyant Tools (Sinclair & Geoffrey, 2016) is an open-source, web-based text reading and analysis environment that uses more than 20 visualization tools to analyze a text corpus, which allows users to investigate patterns of words/concepts and to explore and visualize a large corpus of text systemically—exercises that may be difficult to perform by simply reading. Voyant Tools has been used for research and peer-reviewed publications (Boyle & Hall, 2016). The content analysis implemented in the patents' title list consist in the frequency of terms (i.e., the number of times a word is mentioned in a corpus), their ratio, and the collocation graphs of patent titles. A ratio of terms depicts the changes in the frequency of words included in a corpus where each analyzed group is represented in a vertical column where the highest frequency terms are plotted. The x-axis displays the group titles and the y-axis displays the relative frequencies. A collocation graph is a network of terms with higher frequency and proximity. The keywords are shown in dark gray and the proximity words in light gray.

A permanent link to Mayor's list is available at the following link for the digital version of this chapter (https://goo.gl/KnZjCc) or in the following QR code for the print version.

4.4 Results

The results present the following information: (1) patents by year; (2) coefficient of invention for 1985–1998; (3) type of inventor (i.e., male, female, group, or company); and (4) the geographical location where the invention was developed. The results of the content analyses include (5) the frequency of terms, their ratio and the collocation graphs of patent titles in three time periods (i.e., 1930–1953, 1954–1977, 1978–2000), by region (i.e., Andean and Caribbean) and type of inventor. As an exploratory insight, (6) the patents' classification is also presented.

Results show an overall decrease in the average number of patents granted by year (p/y). In 1930–2000, the average p/y publication was 24.5. When comparing 1930–1979 (26.6 p/y) and 1980–2000 (17.7 p/y), the p/y decreased by −33.4% (see Fig. 4.5).

The invention coefficient (patents per 100,000 inhabitants) shows an upward trend in 1985–1998 with a peak in the mid-1990s (see Fig. 4.6). Because of data availability, the coefficient was constructed for the number of patents granted rather than the number of patent applications.

The publication of patents by type of inventor is disproportionately concentrated. Individual male inventors have been the main players in registered patents (62%), followed by companies (27%), groups of inventors (two or more inventors) (10%), and individual female inventors (1%) (see Fig. 4.7).

As the exercise of invention in Colombia has been essentially an individual enterprise, the geographic origin of patents is diverse. However,

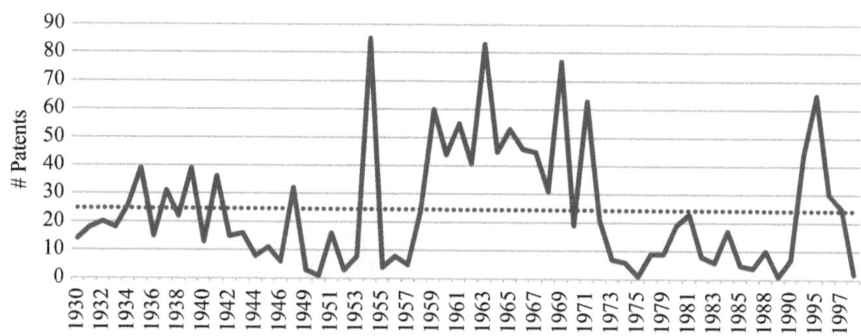

Fig. 4.5 Number of patents granted in Colombia in 1930–2000. Source: The author based on Mayor, 2005

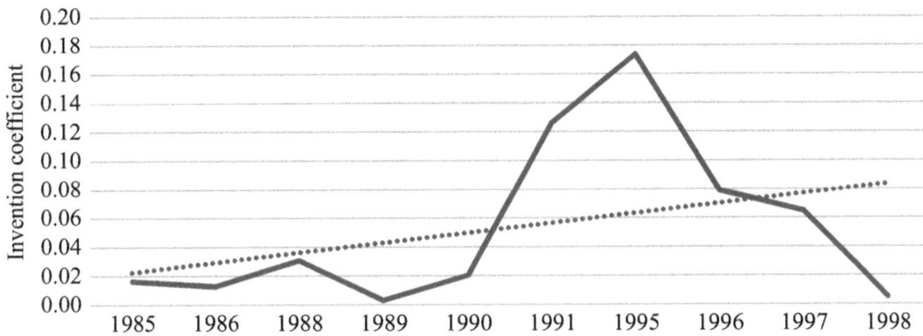

Fig. 4.6 Invention coefficient 1985–1998. Source: The author based on Mayor, 2005, and the National Statistics Department (DANE), 1985–1998

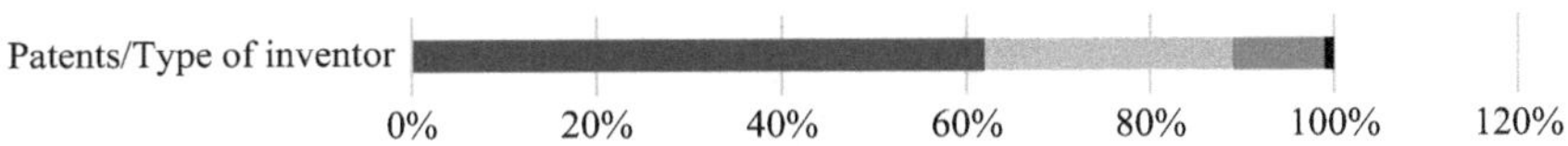

	Patents/Type of inventor
■ Individual male	62%
■ Companies	27%
■ Groups of inventor	10%
■ Individual female	1%

■ Individual male ■ Companies ■ Groups of inventor ■ Individual female

Fig. 4.7 Publication of patents by type of inventor. Source: The author based on Mayor, 2005

over time it has become highly concentrated in larger cities. For instance, a patent from Hoyorrico (Antioquia) can be found in the database. It was granted in 1965 and entitled *Herramientas para extraer y expulsar plantas y envase* (Tools for removing and ejecting plants and containers). Today, more than 50 years later, Hoyorrico remains a small township with less than 1000 inhabitants. Despite these outliers, the geographic origin of 95% of the patents is the main cities: Bogotá (64%), Medellín (15%), Cali (10%), Barranquilla (4%), and Manizales (2%).

Regarding the content analysis of the patents' titles. 9128 words or 56,460 characters with no spaces were analyzed by using Voyant Tools. The top 10 words mentioned were *mejoras* (improvements) (144), *sistema* (system) (125), *máquina* (machine) (95), *aparato* (device) (63),

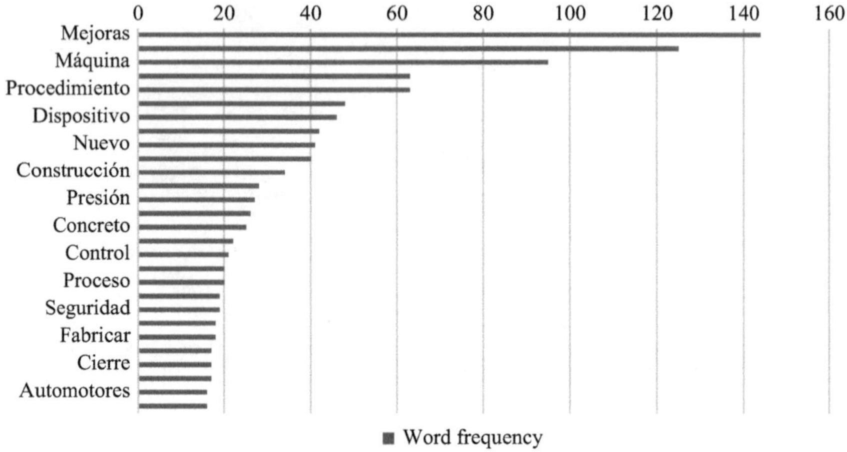

Fig. 4.8 Content analysis of word frequency in patent titles. Source: The author based on Mayor (2005) and analyzed using Voyant Tools

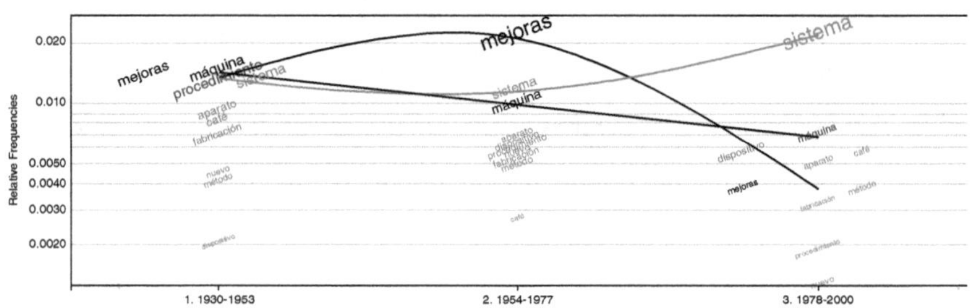

Fig. 4.9 Ratio of the terms in patent names in 1930–1953, 1954–1977, and 1978–2000. Source: The author based on Mayor (2005) and analyzed using Voyant Tools

procedimiento (procedure) (63), *fabricación* (manufacturing) (48), *dispositivo* (device) (46), *café* (coffee) (42), *nuevo* (new) (41), and *método* (method) (see Fig. 4.8).

The ratio of terms in patent titles fluctuated significantly in each study period (see Fig. 4.9). From 1930 to 1953, most patents concerned *máquina* (machines), followed by *mejoras* (improvements) and *sistemas* (systems). The keyword *máquina* was proximate to *café* (coffee), *palma*

Fig. 4.10 Collocation graph of keywords in patent names in 1930–1953. Source: The author based on Mayor (2005), and analyzed using Voyant Tools

(palm), and *desfibrar* (to defibrillate) (see Fig. 4.10). From 1954 to 1977, the ratio of terms shifted. The predominant term was then *mejoras* (improvements), followed by *sistema* (system) and *máquina* (machine). *Mejoras* (improvements) was then proximate to *fabricación* (manufacturing), *construcción* (construction), and *eléctricos* (electrics), while in the previous period *mejoras* (improvements) was proximate to *café* (coffee), *máquinas* (machines), and *descerezadoras* (coffee pulper) (see Fig. 4.11). In 1978–2000, the ratio of terms changed once more. The predominant term became *sistema* (system), followed by *control* (control) and *máquina* (machine). *Mejoras* (improvements) was much further down the ranking. *Sistema* (system), as a keyword, was proximate to *contrucción* (construction), *modular* (modular), *fijación* (fixing), and *agarre* (grip), while in the previous period it was proximate to *eléctrico* (electric), *fabricación* (manufacturing), and *muros* (walls) (see Fig. 4.12).

The majority of patents were developed in cities within the Andean and Caribbean regions. The differences among keywords were not discernible, although a few differences were identified regarding the collocated terms. In the Andean region, the most frequent keywords were *mejoras* (improvements), *sistema* (system), and *máquina* (machine); *mejoras* (improvements) was collocated alongside terms such as *máquina*, *café* (coffee) and

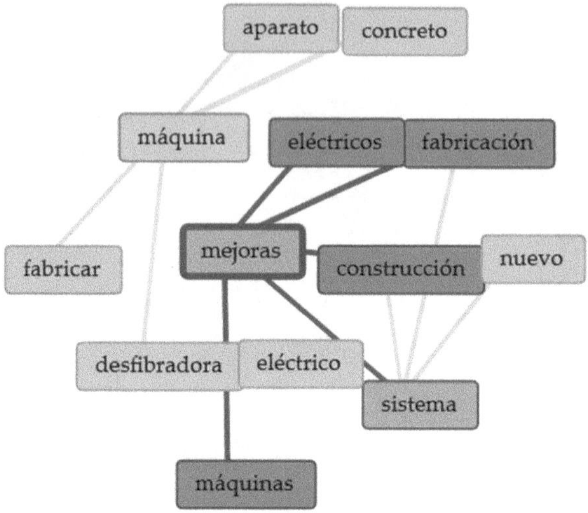

Fig. 4.11 Collocation graph of keywords in patent names in 1954–1977. Source: The author based on Mayor (2005), and analyzed using Voyant Tools

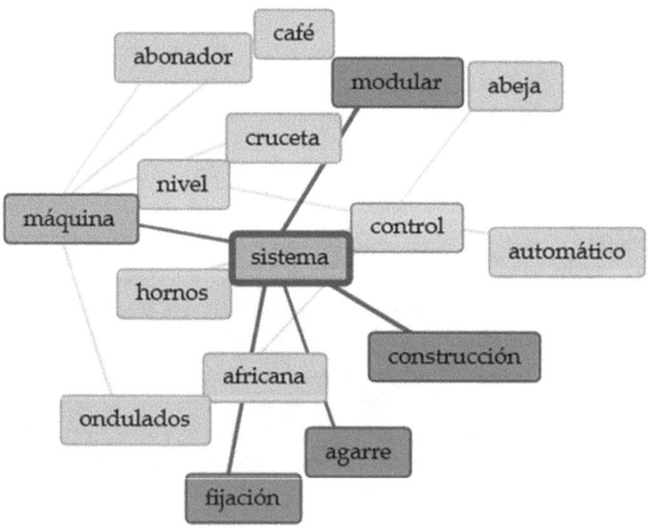

Fig. 4.12 Collocation graph of keywords in patent names in 1978–2000. Source: The author based on Mayor (2005), and analyzed using Voyant Tools

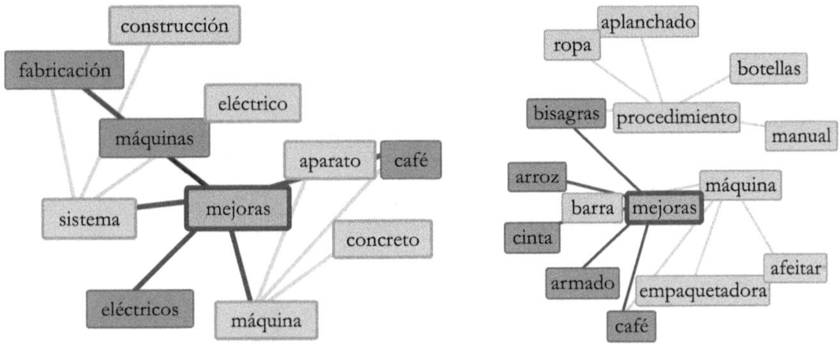

Fig. 4.13 Collocation graph of keywords in patent names granted in cities in the Andean (left) and Caribbean (right) regions. Source: The author based on Mayor (2005), and analyzed using Voyant Tools

fabricación (manufacturing). Otherwise, in the Caribbean region, the most frequent keywords were *máquina* (machine), *mejora* (improvements), and *procedimiento* (procedure); *máquina* (machine) was collocated alongside terms such as *café* (coffee), *afeitar* (to shave), and *empaquetadora* (packing machine) (see Fig. 4.13).

The majority keywords in patent names developed by companies were *mejoras* (improvements), *sistemas* (systems), and *procedimientos* (procedures) (see Fig. 4.14).

In the case of groups of inventors, the most common keywords were *máquina* (machine), *sistemas* (systems), and *mejoras* (improvements) (see Fig. 4.15).

Regarding individual male inventors, the most common keywords were *mejoras* (improvements), *sistema* (system), and *máquina* (machine) (see Fig. 4.16).

Regarding individual female inventors, the most common keywords were *dispositivo* (device), *sistema* (system), and *dispensador* (dispenser) (see Fig. 4.17).

The first edition of the IPC, then the International (i.e., European) Classification of Patents for Invention, was published in 1968. There was no standard classification for patents prior to that date, whether in Colombia or any other country. The IPC establishes five levels of patents' classification: (1) section; (2) class; (3) sub-class; (4) group; and

Fig. 4.14 Collocation graph of keywords in patent titles granted to companies. Source: The author based on Mayor (2005), and analyzed using Voyant Tools

Fig. 4.15 Collocation graph of keywords in patent titles granted to groups of people. Source: The author based on Mayor (2005), and analyzed using Voyant Tools

Fig. 4.16 Collocation graph of keywords in patent titles granted to individual male inventors. Source: The author based on Mayor (2005), and analyzed using Voyant Tools

Fig. 4.17 Collocation graph of keywords in patent titles granted to individual female inventors. Source: The author based on Mayor (2005), and analyzed using Voyant Tools

(5) complete classification symbol (WIPO, 2017a). A section is the higher level of hierarchy of the IPC. As an exploratory insight to classify the patents granted prior to 1968 (i.e., 1039 patents [68%]) and the rest, we classified the patents according to the eight IPC-2017 sections:

- A (section symbol): Human necessities.
- B: Performing operations; transporting.
- C: Chemistry; metallurgy.
- D: Textiles; paper.
- E: Fixed constructions.
- F: Mechanical engineering; lighting; heating; weapons; blasting;
- G: Physics; and
- H: Electricity.

Twenty-eight percent of the patents belong to A section (e.g., *Desceresadora de café* [coffee pulper]), 22% to B (e.g., *Horno especial crematorio para basuras* [Special crematorium for garbage]), 7% to C (e.g., *Desmanchado instantáneo de oxido de hierro* [Iron oxide instant removal]), 4% to D (e.g., *Mejoras en máquina para teñir medias e hilo* [Improvements in machine to dye tights and thread]), 15% to E (e.g., *Máquina portátil para aserrar madera* [Portable wood sawing machine]), 10% to F (e.g., *Semáforo para regular tránsito en ciudades* [Traffic light to regulate traffic in cities]), 6% to G (e.g., *Regla de nomenclatura química* [Chemical nomenclature rule]) and 4% to H (e.g., *Enchufe eléctrico automático* [Automatic electric plug]) (see Fig. 4.18).

4.5 Discussion

One reason why the number of successful patent applications decreased after 1979 was the implementation of a more rigorous legal framework on industrial property (Mayor, 2005). Colombia endorsed the PCT in Washington DC in 1971 (the PCT was amended in 1979). The PCT required Colombian inventors to acknowledge and meet international standards of industrial property. But for a few exceptions (e.g., Parma, Hakim, Barraquer, and Miller) they were far behind the pace of the world leaders. Consider Thomas Alva Edison, who held a world record of 1093 patents: 72% of the total patents produced by Colombian inventors during 70 years (1930–2000) (Encyclopedia Britannica, n.d.). Additionally, the PCT considered *an international search* as a mandatory process for an international patent application, namely (WIPO, 1971):

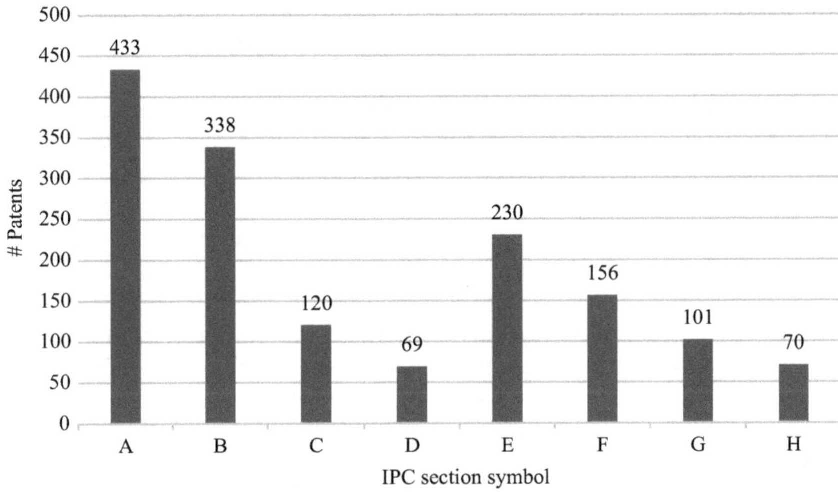

Fig. 4.18 Number of patents by IPC section. Source: The author based on Mayor (2005) and WIPO (2017a)

Article 15—The international search.

(1) *Each international application shall be the subject of international search.*
(2) *The objective of the international search is to discover relevant prior art.*
(3) *International search shall be made on the basis of the claims, with due regard to the description and the drawings (if any).*
(4) *The International Searching Authority referred to in Article 16 shall endeavor to discover as much of the relevant prior art as its facilities permit, and shall, in any case, consult the documentation specified in the Regulations.*

Although patents such as *Caja de cigarrillos y cerillas en un solo paquete* (Pack of cigarettes and matches in one package) were suitable for industrial property protection before the PCT, it is noticeable that similar products were patented with neither a regional nor an international search for similarities. In fact, the most productive years of the study period were 1954 (85 patents), 1963 (83 patents), and 1969 (77 patents), all before 1971. It seems that patenting before the PCT was not as complex as thought, despite the fact that before the PCT Colombia was also the contracting country of several international agreements on industrial property, such as the Paris Convention for the Protection of Industrial Property (1883); the Industrial property agreement with France (1901); the General Inter-American Convention for Trademark and Commercial Protection (1901); the Inter-American Convention on the Legal Regime of Powers of Attorney to be Used Abroad (1940); and the Convention Abolishing the Requirement of Legalization for Foreign Public Documents (1961) (SIC, n.d.-b). Such additions to international agreements did not function as lighthouses for some distant and innovative outlooks to national inventors.

The invention coefficient shows an upward trend in 1985–1998 for two main reasons: (1) a sustained decrease in the population rate and (2) an increase in the patents granted, but only for 1991–1997. The population growth rate between 1985 and 1986 was 4.2% but then decreased to 1.6%, 1.5%, and 1.4% in 1996, 1997, and 1998, respectively (DANE, 1985–1998). Furthermore, the average number of successful patents per year (excluding 1991–1997) was 4.8. From a historical and productivity perspective, this cannot be conceived as an improvement considering that the average number of granted patents per year during the first decade of the study (1930–1940) was 23.1. These inversed trends (i.e., population growth rate stagnation and 1991–1997 as outliers) have helped to encourage the upward trend of the invention coefficient.

Another characteristic of patents in Colombia is the prominence of the individual male inventor. This predominance was considered obsolete among the industrial superpowers, even at the beginning of the twentieth century (Godin, 2017). Individual male inventors have been the backbone of Colombian patenting history, responsible for 62% of the analyzed patents (Mayor, 2005). Without a doubt, some of them introduced revolutionary changes to their fields (e.g., Salomón Hakim and the Hakim Valve changed the field of brain surgery), although the trend among industrial superpowers was heading toward organized research funded by an entrepreneurial State willing to take risks to develop General Purpose Technologies such as electricity, computers, and the Internet (Mazzucato, 2013). According to Maurice Holland (1928a), the director of the division of engineering and industrial research at the US National Research Council, less than 5% of patents that reach the commercial stage are the result of individual inventors in the United States. Holland also added (Holland, 1928b, pp. 4–6):

> *Genius no longer play the leading role in the drama of modern industry… The laboratory has become the adventurer on the frontier of industry… The research worker is a unit in the organization… No single inventor, independent or otherwise, could have developed the transatlantic telephony… This was the product of organized effort.*

A further key point is the gender gap between patent inventors. For each patent produced by an individual female inventor in Colombia in 1930–2000, there were 50 patents produced by individual male inventors (i.e., a total of 19 patents in 70 years). The first patent granted to a woman was in 1932, 22 years before the female vote was established in Colombia. Her name was Constanza Sanín de Díaz and the patent was entitled *Sistema para atraer clientela a comercio al detal* (System to attract customers to retail). Furthermore, global superpowers find themselves in the same situation. In the United States, less than 20% of all patents involved at least one female inventor (Institute for Women's Policy Research [IWPR], 2016). The US Department of Commerce cited the following reasons for the small role played by female inventors: a lack of incentives in Science, Technology, Engineering, and Mathematics (STEM) education and STEM employment to women, few female STEM role models, and strong gender stereotypes. These factors discourage women from pursuing a career in STEM (US Department of Commerce, 2011). In fact, less than 30% of the world's researchers are women (UNESCO Institute for Statistics [UIS], 2015). The absence of women in STEM is a skill gap and unexploited potential that inhibits innovation and economic growth (Inter-American Bank, 2014). A diverse set of policies and incentives has been implemented in developed countries, such as the creation of national committees on women and science, publishing sex-disaggregated statistics, and providing economic incentives to gender-balanced research departments (Inter-American Bank, 2014). Colombia could learn from best practices and adopt suitable policies and incentives according to its historical and institutional context to increase the participation of women in STEM.

Two of the numerous reasons for the dominance of the main cities in patent publication are the urbanization process and urban agglomeration (i.e., ideas). Back in 1968, Rother (1968) estimated that the annual population growth between 1951 and 1964 in Colombia was 3.2%, and that the urban population growth rate during the same period was 2.2 times the growth rate of the whole country. The job market also changed. The number of agriculture-related jobs was decreasing and service-related jobs were increasing in the main cities. In Bogotá, the annual population growth from 1951 to 1964 was 7.0%–7.9%. This urban growth also boosted patent publication in the construction sector: 20.7% of the patents granted between 1961 and 1990 in Bogotá-Cundinamarca were in the construction industry (Mayor, 2005). Additionally, the headquarters and laboratories of the most important companies in the country were located in the capital (e.g., Bavaria, Cementos Samper, and Eterni). However, the urbanization process led to several benefits in terms of wealth creation and innovation. Jones and Romer (2009) argued that innovation is the result of discovering new ideas and the number of new ideas is proportional to the population. Bettencourt, Lobo, and Strumskyc (2007); Bettencourt, Lobo, Strumskyc, and West (2010) found a super-linear effect whereby new patents are granted in larger urban areas, hence producing increasing returns in inventing activity with respect to population size. This trend can still be seen today in Colombia. The five most populous departments are Cundinamarca, Antioquia, Valle del Cauca, Atlántico, and Bolívar (with a total of 25.8 million inhabitants). Consequently, those five departments produced 84% of all patents registered between 2005 and 2014 (OCyT, 2015).

As noted in the content analysis of the patent titles, the descriptive titles of the patents were aimed to improve something already invented; *mejoras* (improvements) was mentioned 144 times. The first patent in the sample aimed to *improve* something that was granted in 1930: *Mejoras en cajitas de fósforos* (Match boxes improvements). Before that, in 1925, the Colombian legal framework developed Law 31, which strictly defined the term "invention patent" (Mayor, 2005, p. 84):

> *Discoveries, improvements or inventions of new industrial products, new ways and new applications of ways to obtain the result of an industrial product.*

"Total novelty" was not at all essential for a patent to be granted. Article 9 stated that invention patents could be granted with no previous novelty or utility assessment, as already required by several international frameworks (Mayor, 2005, p. 86). Therefore, before the PCT in 1971, *mejoras* (improvements) appeared 122 times, almost 84% of the word's total frequency in this study.

The ratio of terms presented provides a detailed insight into patents' keywords and associated terms. The predominant keywords in 1930–1953 were *maquina* (machine) and *mejoras* (improvements), and both were associated with *café* (coffee) and *descerezadoras* (coffee pulper). This could be historically associated with the growth of the coffee industry in Colombia at the beginning of the twentieth century, and the requirement to modernize and *improve* the technologies and *machines* for coffee production and exportation (Bejarano, 1987).

After 1953, the main keyword was *sistema* (system), which was associated with *contrucción* (construction), *modular* (modular), and *muros* (walls). As discussed in the previous sections, it seems that the acceleration of the urbanization process (Rother, 1968) in 1954 also boosted patent publication in related activities (e.g., *sistema de muros a base de columnas, láminas o vigas* [system of walls with columns, sheets or beams]). This was also clear in the regional analysis. As the main cities (i.e., Bogotá, Cali, and Medellín) were located in the Andean region, keywords such as *máquina* (machine) and *sistema* (system) were associated with *concreto* (concrete) and *construcción* (construction).

Among the type of inventors, patents granted to both companies and individual male inventors have *mejoras* (improvements) as the most common keyword. However, in the case of companies, this was mostly related to *construcción* (construction), whereas for individual male inventors it concerned *café* (coffee). This shows the importance of individual inventors in the rapid development of the coffee industry and companies to the urbanization/construction progress. Regarding individual female inventors, one contributed to a nascent and highly sophisticated sector: aviation. In 1968, a patent was granted to María Paz Arce de Luege de La Rosa for the *Dispositivo para transmitir mensajes de avión a tierra* (Device to transmit messages from airplane to land) in the city of Barranquilla.

These findings are in contrast to the patents registered in international offices. While national patenting in 1930–2000 was ruled by individual male inventors and boosted mainly by the coffee industry and the accelerated growth of cities, international patenting in the United States and Europe was led by companies and other types of organizations in several technological areas such as electric digital data processing (Sánchez et al., 2007).

It seems there is a relation between patents' IPC section from 1930 to 2000 and the share of patent application by field of technology today, at least in the top five. Today's top five patent applications by field of technology, excluding "others" category (39.7%), are (WIPO, 2017b): (1) medical technology (9%) related to section A (28%); (2) civil engineering (9%) related to section E; (3) basic materials chemistry (8%) related to section C; (4) food chemistry (6%) also related to section C; and (5) other special machines (6%) related to section F (Fig. 4.19).

As the productive structure of countries is path-dependent, future products generated by a country depend on current capabilities historically accumulated (Hidalgo, Klinger, Barabási, & Hausmann, 2007), in which case Colombia has accumulated a stock of knowledge to keep applying for patents in the technology fields related to that stock (i.e., human necessities and fixed construction, mainly). Nevertheless, as mentioned, these analyses are essentially exploratory.

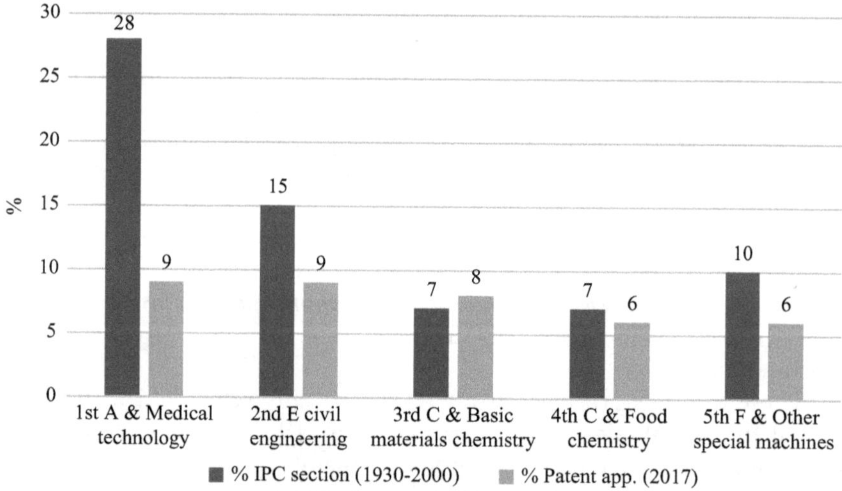

Fig. 4.19 Top five IPC sections from 1930 to 2000 and patent application by field of technology 2017. Source: The author based on Mayor (2005) and WIPO (2017a)

4.6 Conclusions

The wealth of nations is now determined by their capabilities to produce intangible capital, that is, knowledge stock. One of the outputs of this kind of knowledge is patents. Patent publication is unevenly distributed around the world, and those disparities are partly explained by the type of economic and political institutions constructed and adopted, or enforced, by nations throughout history. This is one reason why we need the lenses of history to understand the current situation of any nation.

However, in Colombia, there is very little history-related literature on patents. The findings of Sánchez et al. (2007) showed that the majority of international patents registered between 1968 and 2007 involving at least one Colombian resident/organization, were registered in the United States in technology-related areas, and led by private and other organizations. The rest of the studies did not go beyond the last 15 years. The work of Mayor (2005) enabled the extension of the study period to cover 1858–2000. We focused our analysis on 1930–2000 due to the completeness of sources gathered by Mayor after 1930. Also, we were better able to analyze the main characteristics of Colombian patents, the regions, and the inventions, as well as the evolution of the national legal framework, along with many other aspects grounded in STS studies. Based on that roster of contributions, this study transformed the physical list constructed by Mayor into an open-access dataset to facilitate the diffusion of that research to further studies or replications.

Our analysis discussed the changes in national registered patents publication due to the upgrade of the legal framework to international standards, the upward trend of the invention coefficient due to the stagnation of the population growth rate, and outperforming years of publication. Colombia has been a contracting country of several international agreements on industrial property since 1883. However, the implementation of these standards did not work as an automatic diffusion of the international frontier (i.e., industrial laboratories) to the Colombian lonely inventor. Until 1979, Colombia endorsed the PCT, which forced Colombian inventors to meet international standards. From then on, the total number of patents granted per year decreased by 33.4%.

We also investigated the overwhelming concentration of patents granted to individual male inventors and the obsolescence of the *genius inventor* over *organized research* and *the laboratory* promulgated by the industrial global powers since the beginning of the twentieth century. This could be considered one of the most prominent setbacks in the analysis, considering that for each patent granted to an individual female inventor there were 50 patents granted to individual male inventors. The lack of women in STEM is a serious constraint for innovation and economic growth in developing countries. A new set of policies and incentives should be encouraged to close the gender gap in STEM-related activities, such as the creation of national committees on women and science, publishing sex-disaggregated statistics, or funding national research projects with gender-balanced work teams.

We found an association between the changes to the patents' keywords and major economic changes in the past century: the takeoff of the coffee industry and the urbanization process. The requirements to modernize the production of coffee for internal demand and exportation led to the invention and improvement of machines in the production phase, mostly *descerezadoras* (coffee pulper). On the other hand, the acceleration of the urbanization process since the 1950s boosted patent publication in related activities. The individual inventor was essential to the patents' publication in the coffee sector, while for the construction sector it was the companies. Considering the growth of cities, we also contrasted these findings with the relationship among urban agglomeration, ideas, and innovation, and the predominance of patent publication in Colombian main cities.

Half of the patents analyzed belong to the IPC A and B sections (i.e., human necessities and performing operations; transporting). Considering the path-dependence of the evolution of national productive structures, there is a relation between both sections mentioned, which dominate the historical patent production landscape and today's top five patent applications by field of technology, such as medical technology, civil engineering, basic materials chemistry, and food chemistry.

The limitations of this study are related to the actualization and consolidation of Mayor's list with the SIC database, and impact measurements of those patents. As the SIC patents database is constantly updated, it is feasible that patents listed in the SIC database could not be found in Mayor's list and vice versa. Mayor's patents list contains no information on patents' citations, as a measure of technological impact (Trajtenberg, 1990). Further studies could create one broader dataset and measure the citations of those patents to develop new technologies. Furthermore, a theoretical and methodological framework such as economic geography could also provide additional insights into the effects of geographical conditions (e.g., weather conditions, proximity, or regions connected by railways) in patent publication at a regional level in Colombia.

Acknowledgments The author would like to thank Daniel García Rojas and William Andrés Muñoz for their research assistance.

References

Acemoglu, D., & Robinson, J. (2013). *Why Nations Fail: The Origins of Power, Prosperity, and Poverty*. New York: Crown Publishers.

Acemoglu, D., García-Jimeno, C., & Robinson, J. (2012). Finding Eldorado: Slavery and Long-run Development in Colombia. *Journal of Comparative Economics, 40*(4), 534–564.

Aguero, C. (2017). Collaborative Networks and Patent Production in Andean Community of Nations Universities (UCANS), 2005–2015. *Revista Española de Documentación Científica, 40*(2), e172.

Bejarano, J. (1987). El despegue cafetero (1990–1928). In J. Ocampo (Ed.), *Historia Económica de Colombia*. Retrieved from https://goo.gl/CcLMxN

Bettencourt, L., Lobo, J., & Strumskyc, D. (2007). Invention in the City: Increasing Returns to Patenting as Scaling Function of Metropolitan Size. *Research Policy, 36*(1), 107–120.

Bettencourt, L., Lobo, J., Strumskyc, D., & West, G. (2010). Urban Scaling and Its Deviations: Revealing the Structure of Wealth, Innovation and Crime Across Cities. *PlosOne, 5*(11), e13541.

Boyle, M., & Hall, C. (2016). Teaching Don Quixote in the Digital Age: Page and Screen, Visual and Tactile. *Hispania, 99*(4), 600–614.

Campos, E., & Campos, C. (2014). Biotechnological Overview of Latin America from the Perspective of Immunological Patents. *Revista Cubana de Información en Ciencias de la Salud, 25*(2), 172–182.

Del Socorro López Gómez, M., Simón, R. S., Gómez, F. C., & Tobón, C. G. (2009). The Processes Involved in a Management Model for University Patents. *Ingenieria e Investigacion, 29*(2), 135–141.

Departamento Administrativo Nacional de Estadística (DANE). (1985–1998). *Proyección de población*. Retrieved from https://goo.gl/wgEJbz

Enciclopedia Britannica. (n.d.). *Thomas Alva Edison, Biography*. Retrieved from https://goo.gl/gurRk9

Godin, B. (2017). *Models of Innovation: The History of an Idea*. London: The MIT Press.

Hall, B., Jaffe, A., & Trajtenberg, M. (2005). Market Value and Patent Citations. *RAND Journal of Economics, 36*(1), 16–38.

Harari, Y. (2015). *Homo Deus: A Brief History of Tomorrow*. London: Harvill Secker.

Hasan, I., & Tucci, C. (2010). The Innovation–economic Growth Nexus: Global Evidence. *Research Policy, 39*(10), 1264–1276.

Hidalgo, C., Klinger, B., Barabási, A., & Hausmann, R. (2007). The Product Space Conditions the Development of Nations. *Science, 317*(5837), 485–487.

Holland, M. (1928a). Research, Science, and Invention. In F. Wile (Ed.), *A Century of Industrial Progress* (pp. 312–334). New York: Doubleday, Doran.
Holland, M. (1928b). *Industrial Explorers*. New York: Harper.
Hu, A., & Png, I. (2010). *Patent Rights and Economic Growth: Evidence from Cross-Country Panels of Manufacturing Industries*. Retrieved from https://goo.gl/Am5Gp8
Institute for Women's Policy Research (IWPR). (2016). *Annual Report*. Retrieved from http://bit.ly/2kKYjtU
Inter-American Bank. (2014). *Women in Science and Technology: What Does the Literature Say?* Washington: Inter-American Bank.
Jaffe, A., & Trajtenberg, M. (1996). Flows of Knowledge from Universities and Federal Laboratories: Modeling the Flow of Patent Citations Over Time and Across Institutional and Geographic Boundaries. *Proceedings of the National Academy of Sciences of the United States of America, 93*(23), 12671–12677.
Jones, C., & Romer, P. (2009). *The New Kaldor Facts: Ideas, Institutions, Population, and Human Capital*. Massachusetts: National Bureau of Economic Research.
Mayor, A. (2005). *Inventos y patentes en Colombia 1930–2000: De los límites de las herramientas a las fronteras del conocimiento*. Medellín: Instituto Técnico Metropolitano.
Mazzucato, M. (2013). *The Entrepreneurial State: Debunking Public vs. Private Sector Myths*. New York: Public Affairs.
Munafò, M., & Smith, G. (2018). Robust Research Needs Many Lines of Evidence. *Nature, 553*, 399–401.
North, D. (1989). Institutions and Economic Growth: An Historical Introduction. *World Development, 17*(9), 1319–1332.
Observatorio Colombiano de Ciencia y Tecnología (OCyT). (2004–2015). *Indicadores de ciencia y tecnología – Colombia*. Bogotá: OCyT.
Observatorio Colombiano de Ciencia y Tecnología (OCyT). (2015). *Indicadores de ciencia y tecnología, Colombia*. Bogotá: OCyT.
OCyT. (n.d.). *Historia*. Retrieved from https://goo.gl/vN3LYN
Open Science Collaboration. (2015). Estimating the Reproducibility of Psychological Science. *Science, 349*(6251), 943.
Rennie, R. (2002). History and Policy-making. *International Social Science Journal, 50*(156), 289–301.
Robledo, J., & Saavedra, J. (2016). Patents and Economic Growth: ¿Innovation of Residents and Non-residents? *Desarrollo y Sociedad, 2016*(76), 243–272.
Rother, H. (1968). El proceso de urbanización en Colombia. *Revista de la Universidad Nacional (1944–1992), 1*, 191–223.
Sánchez, J., Medina, J., & León, A. (2007). The International Publication of Patents by Colombian Organizations and Inventors. *Cuadernos de Economía, 26*(47), 247–270.
Sinclair, S., & Geoffrey R. (2016). *Voyant Tools*. Retrieved from http://voyant-tools.org/

Superintendencia de Industria y Comercio (SIC). (n.d.-a). *Historia*. Retrieved from https://goo.gl/nTTjM9

Superintendencia de Industria y Comercio (SIC). (n.d.-b). *Compendio de normas*. Retrieved from https://goo.gl/xcc2Gk

The World Bank. (2011). *The Changing Wealth of Nations: Measuring Sustainable Development in the New Millennium*. Washington: The World Bank.

Trajtenberg, M. (1990). A Penny for Your Quotes: Patent Citations and the Value of Innovations. *The Rand Journal of Economics, 21*(1), 172–187.

UNESCO Institute for Statistics. (2015). *STEM and Gender Advancement*. Retrieved from https://goo.gl/AnL66n

US Department of Commerce. (2011). *Women in STEM: A Gender Gap to Innovation*. Retrieved from https://goo.gl/kAtGaV

WIPO. (1971). *Patent Cooperation Treaty (PCT)*. Retrieved from https://goo.gl/64Tdu4

WIPO. (2015). IP Statistics Data Center. Retrieved from https://www.wipo.int/ipstats/en/

WIPO. (2017a). *Guide to the International Patent Classification*. Retrieved from https://goo.gl/PSKvLm

WIPO. (2017b). *Statistical Country Profiles: Colombia*. Retrieved from https://goo.gl/UZ9BXj

World Intellectual Property Organization (WIPO). (n.d.). *What Is a Patent?* Retrieved from https://goo.gl/DXz24h

CHAPTER 5

Open Access Policies in Science to Promote Science, Technology and Innovation in Emerging Economies. The Case of Colombia

Karen Isabel Cabrera Peña and Patricia Beatriz Márquez Rodríguez

5.1 Introduction

In general, one of the key questions that every country faces is related to the creation of strategies to increase its scientific, innovative and technological capacity in order to improve its social and economic welfare. Taking this into account, over recent years there has been growing interest in the movement of Open Access in Science or Open Science, as a means to increase the effectiveness and efficiency of the scientific and research efforts (the Organisation for Economic Co-operation and Development, OECD, 2015).

K. I. Cabrera Peña (✉)
Departamento de Derecho, Universidad del Norte, Barranquilla, Colombia
e-mail: cabrerak@uninorte.edu.co

P. B. Márquez Rodríguez
Escuela de Negocios, Universidad del Norte, Barranquilla, Colombia
e-mail: patriciarodriguez@uninorte.edu.co

C. I. Pardo Martínez et al. (eds.), *Analysis of Science, Technology, and Innovation in Emerging Economies*,
https://doi.org/10.1007/978-3-030-13578-2_5

Open Access was born as a way to allow free access to readers through digital and online versions of scientific information (Cabrera, 2015). However, the scope of this concept has reached complementary elements including the public access to research data, tools, publications and results, expanding the possibility to share the knowledge and the impact of the whole research process. According to this, and following Levin, Leonelli, Weckowska, Castle, and Dupré (2016), in this chapter we consider Open Access in Science broadly, including the access to, dissemination of and re-use of publications, data, materials and methods.

The notion behind Open Science is that "Economists consider scientific knowledge generated by public research as a public good, which means that everyone can use that knowledge at no additional cost once it is made public, generating higher social returns" (OECD, 2015, p. 11). In that sense, the public access to science has been linked to a greater impact on business, innovation and society, increasing the direct effects and the spillovers arising from science and research (OECD, 2015).

Although the origins of the concept can be traced to several decades ago, the term was introduced recently by Paul David in 2003 (OECD, 2015) and it has had a renewed interest due to the massive development of Information and Communication Technologies (ICT), which in turn play an important role in the promotion and effective implementation of the sharing of scientific knowledge.

Taking this into account, during the last few years there have been several initiatives at national level to exploit the benefits of Open Access in science, technology and innovation (STI) performances; however, these efforts are heterogeneous and require the development of policies adapted to the context of developed as well as developing countries.

In this chapter, we will analyze the case of Colombia as a developing country, in order to identify the fundamental elements of policy which enable the alignment of the expected outputs in STI with the efforts around Open Science initiatives, and compare these elements with some of the most relevant experiences in other countries.

5.1.1 Literature Review

One of the most relevant precedents in the field of Open Access in Science is the Budapest Open Initiative, which was created in 2001 as a document to consider the need to adapt scientific communication to the contemporary context and to the impact of Internet and ICT. The purpose of this declaration was to establish a set of recommendations as a means to encourage the development and joint action of Open Access policies in higher education institutions and in funding agencies; the open licensing of scholarly works; the development of infrastructure such as Open Access repositories; and the creation of standards of professional conduct for Open Access publishing (Budapest Open Access Initiative, 2002).

As a main result, the Budapest Open Access Initiative (2002) introduced two strategies. The first is self-archiving, which means that authors can upload a copy of their peer-reviewed publications in Open Access repositories, which have been called Green Road. The second strategy was the development of Open Access journals, called Gold Road, which implies the emergence of new business models that replace charging for access to the current restricted access journals.

Since the Budapest Open Access Initiative, there have been additional efforts to strengthen the Open Access movement such as the Bethesda Statement and Berlin Declaration in 2003, among several others (Cabrera, 2014). According to Costa and Leite (2016) these documents are generated by researchers in different geographical locations all over the world, in order to explain the importance of the public access to scientific information and configured commitments and strategies to achieve the goal.

After all these international efforts, the multidisciplinary nature of Open Access in Science has been recognized, including additional elements to the Open Access at all stages of research, and therefore it requires the participation and action of different actors like governments, research institutions, journals, universities and communities, among others. Thus, parallel to these initiatives, the governments of the countries involved in the raising of the Open Science movement have developed legal sources as well as public policies to regulate the access to the scientific and academic information (Cabrera, 2014).

The OECD is one of the international entities which have developed a common framework of knowledge for its members (mostly developed countries), including principles and guidelines to promote broad policy recommendations with the purposes of developing publicly accessible scientific research and knowledge (OECD, 2017).

The OECD members have made efforts to adapt legal frameworks and implement policy initiatives in order to encourage greater openness in science. The final goal is to promote better results in terms of STI by spreading the effect of knowledge through their national science systems.

Another relevant effort at the transnational level is the case of the European Union, which considers Open Science as one of their three strategic priorities according to the research and innovation policy and has developed an Open Science Policy Agenda for Europe. This policy agenda determines policy actions in the following areas: to foster Open Science, remove barriers to Open Science, develop research infrastructures for Open Science, mainstream Open Access to research results and embed Open Science in society (European Commission, 2017a).

In spite of the consensus about the positive effects of Open Access in Science, there are scarce and heterogeneous initiatives at the national level, both for developed and developing countries, related to the measurement of impact on national systems of STI. This could be explained in part by the lack of consistent data and information on Open Access variables over time, with which to compare the impact of the scientific performance of the countries.

To face this obstacle, the European Union has designed the Open Science Monitor as a way to evaluate the development of Open Science initiatives in Europe over time, as well as to compare the performance of these activities between countries and scientific disciplines. In order to monitor the results of the implementation of Open Science, they have defined four core areas to monitor: Open Access, Open Research Data, Open Scholarly Communication and Citizen Science (European Commission, 2017b).

More recently, the OECD has introduced a set of experimental indicators to the OECD Science, Technology and Industry Scoreboard 2017: Open Access to scientific documents and highly cited documents by Open Access status. These indicators were introduced to address the Open Access and Open Science results and also to provide policy makers a way to assess the social costs and benefits of Open Access (OECD, 2017).

However, the results of these metrics are still at the experimental level and highly focused on the citation of documents, which were not able to capture the usefulness and relevance of the research, especially in the case of non-academic actors that also could be potential users of scientific knowledge (OECD, 2017), and in turn, significant players in the national systems of STI.

The assessment of the economic effects of Open Access, as another impact associated with these phenomena, is also an emergent topic of great interest but also few studies. Houghton and Sheehan (2009) have calculated the returns to R&D of increasing accessibility to research findings in Australia, concluding that the advantages in terms of accessibility generate a return of approximately AUD nine billion over 20 years. In addition, similar studies (Beagrie & Houghton, 2014; Houghton, 2009; Houghton, Rasmussen, & Sheehan, 2010) have shown that the economic effect of Open Science results would be significantly higher than the estimated cost of implementation (OECD, 2015).

These results imply the need for further development and monitoring of complementary tools in order to keep on track the actual impact of Open Science, as well as the deep understanding about the policy considerations in order to promote the public sharing of scientific outputs which involve a multidisciplinary approach including institutional and legal, among other considerations.

5.2 The Case of Colombia and Latin America

In the case of developing countries, specifically Latin America as the target of this chapter, the importance of Open Access to increasing outputs in science and technology performances has also been highlighted, in order to reduce the effects of the current obstacles in accessing international sources of top scientific publications and low national investments in science and technology (Chan & Costa, 2005). These obstacles have contributed to generating research outputs characterized by little impact and isolation, which in turn lead to low levels of development and scientific outputs in the developing world (Chan & Costa, 2005).

According to the results of the Network for Science and Technology Indicators–Ibero-American and Inter-American (RICYT)–there has been an important rise of the science efforts in the region, represented by an increase of 37% of scientific publications in the 2010–2015 period. Taking into account the results of the Scopus scientific database, however, these results do not reflect a similar impact on the technological development of the region (RICYT, 2017).

These results show that scientific knowledge—mainly represented by publications generated in universities of this region—has not effectively reached non-academic actors, producing a scarce contribution to the innovation performance of firms (RICYT, 2017).

Although the data and information specifically regarding Open Access in Science is almost non-existent, this finding suggests the importance of the design of effective policies in order to ensure the greater impact of knowledge in developing countries, with the introduction of new and renewed mechanisms to improve the access to top scientific publications as well as the access and exploitation of research outputs.

According to Costa and Leite (2016), the Latin American Region has developed a strong propensity to develop Open Access mechanisms, especially in the case of Brazil where Gold Access has clearly developed in recent years (OECD, 2017). However, the developments of policies around the implementation of Open Access and Open Science have been more difficult, and there has not been a clear path to take advantage of the potential of Open Access to spread the results of scientific knowledge, especially beyond academic actors.

In the contexts of developing countries, the design and implementation of an effective public policy on Open Access and Open Science is especially important because it could facilitate the exploitation of research outputs, and also improve the productivity of science systems through the greater access to scientific inputs and outputs (OECD, 2015). In turn, it is important to recognize that Open Science policies should be principle-based but adapted to local realities, according to the context and the actors participating in the different mechanisms created to enhance the access to scientific knowledge (OECD, 2015).

For the purpose of this chapter we will analyze the case of Colombia and the degree of development of public policies related to Open Access in Science, and we will also compare the results of this analysis with the experience of other countries with more significant results in this field, in order to establish the main policy considerations for an emerging economy.

5.3 Methods

This qualitative study proposes two perspectives on the study of the Open Access policies in Science in order to promote STI in emerging economies, using Colombia as an example.

On one hand, this chapter proposes an exploratory and comparative analysis of the Open Access policies associated with science, taking into account four representative national and transnational efforts on this topic. The goal of this analysis was to gain understanding about the policy implications around the implementation of Open Science, especially those policies that could be incorporated in countries with emerging economies like Colombia.

For this first analysis, a specialized documentary revision was made using articles and documents of different entities (some of them accessed via Internet) about previous results in the field, as well as documents on public policies on Open Access in order to identify which countries have led the efforts on this topic. According to the purpose of this study, the efforts that represent global trends are the United Kingdom, the European Union, Australia and Argentina.

On the other hand, a legal review of the National Council of Political and Economic Policies (CONPES) documents was made to assess the public policies on STI, in order to evaluate the association of these policies with Open Access in Science in Colombia. According to the findings of this analysis, a set of policy areas to operationalize Open Science is presented, which can be extrapolated to emerging contexts like Colombia and could address difficulties and obstacles identified by previous national and transnational experiences.

5.4 Results

5.4.1 *Countries with Open Access Policies Associated with STI*

According to the objectives of this chapter, a review of some of the main efforts at the national level addressing the policies of Open Access has been conducted: the European Union, Australia, the United Kingdom and Argentina. The results of the comparison of these policies are summarized in Table 5.1. The main findings show the need for integrated efforts due to the multidisciplinary nature of this topic and they enable us to identify the areas to be addressed with the creation of specific policies.

The most relevant strengths of each case analyzed are compiled as follow:

- In the case of the European Union, it is important to highlight the development of the policy linked particularly to the research and innovation program. In that sense, Open Access has been declared as a general principle and obligatory in order to use the resources of the EU research program.
- Since the early 2000s, Australia has been a pioneer in Open Access, developing policies focused on the increase of the returns of investments in science and research, as well as on the efficiency of the research process by maximizing the value and the impact of the research assets for scientists and the nation in the long term.

Table 5.1 Countries with Open Access policies associated with STI

Country	*Open Access Policy*	*Description*	*STI Connection*	*Difficulties*
European Union	Horizon 2020: EU research and innovation program (2014)	Open access is a general principle. Open access is obligatory for scientific publications with program resources (green or gold road). In open access to research data, the information collected must be open but it must necessarily have a data management plan in order to stipulate the way to make it accessible and to exploit (re-use) the information (European Commission, n.d.)	Currently, there is no study to measure the impact on STI, but the Horizon 2020 Commission intends to improve the quality of previous Open Data results, gaining efficiency with collaboration and avoiding duplication in results and consequently obtaining innovation outputs (European Commission, n.d.). Also, a platform to monitor four core areas of Open Science has been developed in order to collect data in order to measure the effectiveness of the implemented policies.	
	European Open Science Cloud (EOSC) Declaration (2017)	This EOSC develops Open Science and open innovation in Europe. Two important topics to highlight: All researchers in Europe must have access to an open-by-default, efficient and cross-disciplinary research data environment supported by FAIR data principles. Open access must be the default setting for all results of publicly funded research in Europe. The EOSC must be underpinned by minimal and rigorous global standards for open research data, as well as standards for EOSC-based services for collaboration through the EOSC. These standards must combine long-term sustainability with optimal freedom of local implementation (EOSC Declaration, 2017).		

Australia	Australian National Data Services (ANDS) (2008) National Science Statement (2017)	ANDS' core purpose is to make Australia's research data assets more valuable for researchers, research institutions and the nation. It helps to store and manage Australia's research data, and it makes it discoverable and accessible for validation and re-use through open access policy (Australian National Data Services, n.d.). The statement articulates the importance that the Australian government places on having a strong and stable science system and its recognition of the long-term nature of science. In order to maximize the opportunities for all Australians in science progress, government uses Open Access in order to improve the return on investment in science through increased use, re-use and impact of scientific findings (Australian Government, 2017).	In 2014, it was the 15th country with the highest investment in science and research in the world. From 2015 to 2016, US$145 billion was invested, corresponding to 11% of the GDP. According to a report, scientific and technological initiatives in Australia would be contributing more than US$300 billion each year to the country's economy, which represents more than 22% of GDP in 2014 (Australian Academy of Science, 2014).	Some researchers have considered that the data they collect belongs to them and they have refused to share it, especially individual researchers. There are doubts about the data that should be shared and that with a privacy reservation (Picasso & Phelan, 2014).

(*continued*)

Table 5.1 (continued)

Country	*Open Access Policy*	*Description*	*STI Connection*	*Difficulties*
United Kingdom	Open Data White Paper (2012)	Some of the objectives described are easier-to-access public data, a more standardized format and the presumption of publishing most of the data. In addition, it stipulates responsibilities for public bodies which include publishing data in open, standardized formats and maintaining and publishing inventories (Lasthiotakis, Kretz, & Creso, 2015).	Open Access policy facilitated greater research output and productivity, as researchers mutually benefited from increased expertise and research capacity, and they developed ties that led to lasting collaborations. For others, cooperation provided tools and platforms in order to improve the community of researchers, increasing the reproducibility of research, preventing the duplication of effort or loss of knowledge, and ultimately leading to more rigorous results and methods (Levin et al., 2016)	The policy has been criticized because it benefits the publishing business, so England would spend considerable sums of money to maintain the model of publishers instead of supporting the Green Road that forces us to question and rethink the business (Cabrera, 2014).
	Finch Report (2012)	This report is a continuation of the science and technology Committee's 2004 report, "free for all?", evidencing the lack of access to published research findings and the academic publishing market as issues which need to be addressed. As a result, this report proposed to minimize restrictions on the rights of use and re-use, especially for non-commercial purposes, and on the ability to use the attested tools and services in order to organize and manipulate text and other content. To get this reaction it is necessary to strengthen the gold road and propose new rules in the green road in order to ensure the viability for the repositories in the market (UK Parliament, n.d.).		
	Research Councils UK (RCUK) policy on open access	RCUK is a government organization that works to optimize the ways in which the research councils work together in order to enhance the impact of UK research, training and knowledge transfer and to be recognized by academia, business and government for excellence in research sponsorship (Research Council UK, n.d.-a). To accomplish these objectives, RCUK has an open access policy that establishes the following: • it applies only to the publication of peer-reviewed research articles. • it supports gold and green roads to open access, but has a preference for immediate open access with the maximum opportunity for re-use. • research will be available through a block grant awarded directly to research organizations (Research Council UK, n.d.-b).		

Argentina	Digital repository law. 26.899 Law (2013)	Argentina is the second country in Latin America with a law on Digital Repository. This is addressed exclusively to the development and implementation of Green Road in Open Access and the conformation of institutional repositories on Open Data in order to deposit the scientific-technological production financed —total or partially—with public resources (Cabrera, 2015)
	National System of digital repository	National System of digital repository is an interoperable network of digital repositories with establishment of policies, standards and common protocols. The model of the repository stated that users can, for free, read, download, copy, distribute, print, search or link the complete texts of scientific articles, and use them for legitimate purposes linked to scientific research, education or the management of public policies, without economic, legal or technical barriers other than those that imply the internet itself (Ministerio de Ciencia, Tecnología e Innovación Productiva, n.d.).

- The United Kingdom approach has stressed on the collaborative nature of Open Access in Science. In that sense, the emphasis of this policy has included the enhancement of the availability of scientific outputs, decreasing the time to access after publications, with the final goal of expanding their use.
- As a country with contextual factors that are similar to Colombia, Argentina has accentuated the importance of institutional efforts in order to start the operationalization of Open Access in Science. In the same way, as a lesson learned in the Argentinian case, the provision of infrastructure and the complementarity between these two components, Green and Gold Road, are vital for the effective dissemination of scientific knowledge.

5.4.2 Colombian Legal Framework on STI and Its Relation to Open Access

By virtue of the fact that Colombia does not have an Open Access policy, this section aims to study STI policies chronologically and their relation to advances in Open Access in this country.

All national-level policies of Colombia are contained in CONPES documents. CONPES documents may be defined as official papers that incorporate studies and approve the policies on economic and social development in the country (Dirección Nacional de Planeación, n.d.). In the case of the national policies of STI, CONPES carries out the analysis, diagnosis and goals on this topic (Plata & Cabrera, 2011).

One of the first Colombian policies in STI was the CONPES 1640 of 1980, which included the National Integration Plan: Science and technology policy. This plan exposed the need to use science and technology in order to achieve the welfare of the Colombian society and, particularly, to transform or make the country productive and competitive in the international markets (Dirección Nacional de Planeación, 1980).

The purposes of this policy were:

1. To increase science and technology investment because, at that time, scientific development investment was 0.05% of total GDP whereas in other countries it was between 0.70% and 1% (Organization of American States, 1972);
2. To strengthen the productive sector through technology because there was a feeble relationship between production activities and technology-scientific activities;

3. To establish programs to strengthen the national science and technology infrastructure through university research and specialized centers;
4. To encourage technology transfer and dissemination in the productive sector, abroad and with national sources in order to adapt this technology to the needs and conditions of the country (Dirección Nacional de Planeación, 1980).

Regarding point 4, about transfer and dissemination of technology, this CONPES pointed out the dissemination of information related to scientific-technological knowledge through the National Information System as a priority, in order to facilitate the use of existing information in areas such as agriculture, energy and mineral resources (Dirección Nacional de Planeación, 1980). Despite the fact that it was a good initiative, this plan did not specify what kind of information should be found in the system, how this information had to be provided and if this information was protected under any license or used by Open Access.

After some years, in 1991, the CONPES 2540 was issued in order to continue the development of the science and technology policy. According to the same document, Colombia lags behind in incorporating innovative production processes with a real impact on the country's development, taking into account the low results of granted patents per capita. However the investment in science and technology increased to 0.1% of total GDP (Dirección Nacional de Planeación, 1991). As can be seen, the protection and legal registration of creations derived from patents was considered as a measure of economic progress.

Regarding this CONPES, it must be emphasized that it was divided into several components such as economic openness in order to accelerate innovation processes and stimulate the relation between productive sectors and research networks through the consolidation of research networks. Another component is the legal flexibility that is expected as a result of processes of democratization of information and, finally, the decentralization of activities resulting in knowledge development through national programs that target regional needs.

On the topic of legal flexibility, the Plan specifically recommended the modernization of the Intellectual Property System in order to ensure the registration of creations according to the different kinds of legal protection. In this sense, it was indicated that the modernization of the Patent Law System would promote economic openness of the country (Dirección Nacional de Planeación, 1980).

As a result of this CONPES, the investment in STI reached around 0.2% of GDP, investment programs and international technical cooperation was 0.3%, and 45% of the state investments were concentrated in the agricultural sector. However, this rate was still very low considering the economic dynamism of the time, which is proof that the measures were not effective enough to reach the proposed goals.

In 1994, considering the minimal progress in technology and innovation, the CONPES 2739 was issued under the principles of the 1991 Political Constitution of Colombia. A public policy was formulated in order to promote technology, science and innovation, and increase competitiveness of the national productive sector in the context of internationalization of the economy (Plata & Cabrera, 2011). Four fundamental strategies were implemented: first, developing and strengthening the national capacity in terms of science and technology; second, creating competitive conditions in the national productive sector; third, reinforcing the capacity to improve social services and creating knowledge regarding the social reality of the country; fourth, generating and implementing scientific knowledge to Colombian culture through a teaching, dissemination and popularization program (Dirección Nacional de Planeación, 1994).

In order to support research on strategic subjects for social and economic development in Colombia, this CONPES intended to, among other things, facilitate access to new knowledge and promote technology transfer. For that purpose, it was again recommended to strengthen and accelerate the registration of industrial property both national and foreign, promoting greater technology flows, and training on political and intellectual management topics, control systems, and analysis and laboratory tests. Once more, the results were not the best, and in 1996 Colombia was among those countries with the fewest scientific publications in Latin America, where only 1% of the population had Masters and PhD degrees and GDP investment in STI was less than 0.4% (Presidencia de la República de Colombia, 1996).

It is important to highlight that this CONPES was the first one addressing innovation, competitiveness and technological development subjects based on access to information using measures different from the National Information System. In this sense, it was encouraged to develop the use of computers, Internet and software with educational purposes, as well as the creation of information systems in order to have national and international networks dealing with relevant topics for the development of the country. Unfortunately, this CONPES did not specify how the information must be shared, and therefore, it opened up the possibility that research and data might not be shared and used for collaborative work.

Along with the need to strengthen the generation, management and use of knowledge, the 3080 CONPES of the year 2000 provided strategies such as promoting innovation in the productive sector, improving acquired technology and technical consulting for companies, funding STI activities in order to collect data which increases national investment in these topics, and promoting the appropriation of knowledge as well as the use of media (Plata & Cabrera, 2011).

To achieve these goals, as one the actions proposed in the National Plan belonging to the National System of Science and Technology, the Agenda for connectivity was created in the Ministry of Communications to use information technologies for the economic, political and social development of the country. This agenda set goals such as promoting the National University Network and the Better Projects Bank in different disciplines of knowledge, and implementing the Virtual Library concept (Dirección Nacional de Planeación, 2000).

In 2007, the Connectivity Agenda and information networks promotion were materialized in the National Network of Research and Education (known as RENATA), which connects, articulates and integrates the actors of the National System of Information (currently known as the National System of Science, Technology and Innovation). This network could be relevant to Open Access implementation because it provides services, information, tools and technological infrastructure through communication and collaboration between its members so as to contribute to the improvement, productivity and competitiveness of the science and academy of the country (RENATA, n.d.).

In addition, with the operation of RENATA, some of the proposals regarding the strategy of innovation and technological development were achieved and the National System of Information was strengthened because all the institutions became integrated (Dirección Nacional de Planeación, 2000). Also, the integration of academic and productive sectors showed significant progress; hence, there was certainty about the number of ongoing projects, and the number of registered patents, companies and laboratories with quality certificates (Dirección Nacional de Planeación, 2000). However, these were improvements; once again, the policy did not indicate what type of information must be provided, who is required to do so, or what legal standards must be applied.

Besides RENATA, it is important to mention LAReferencia, which started as Biblioteca Digital Colombiana and connects the institutional archives of the country. This interoperable platform of Open Access integrates scientific articles and masters and doctoral theses, as well as institutions of the different countries that are part of it (LAReferencia., n.d.). Even though it is supported by the Ministry of Education and the Ministry of ICT, the network has experienced advances thanks to institutions, mostly private, which are also part of it (Uribe-Tirado, 2016).

The CONPES 3080 recognized knowledge as a public good and thus that it should not be restricted to some regions and communities in the country. Though the objective is public communication and popularization of science and technology, the document stated that it is necessary to develop strategies in order to market STI, promote mechanisms of appropriation and private use of knowledge, and generate a research culture for a specific user (Dirección Nacional de Planeación, 2000). In that sense, it delivered a contradictory message on which strategies must be implemented by innovators in this field; it is not clear if the information had to be released or restricted.

Finally, through the Colombian Observatory of Science and Technology, the examination of the National System of Science and Technology was requested, as well as the indicators that determine the situation of the industrial and commercial sectors of the country (Dirección Nacional de Planeación, 2000).

There are some facts, as a result of this policy, which are important to point out in order to have a clear understanding of the evolution of the STI policies in Colombia:

1. According to the Science and Technology Observatory, in 2006 Colombia's investment in R&D was 0.18% of GDP and in 2007 it was 0.41% (Colciencias, 2010).
2. In 2009 the country had 84 doctorate programs, 52 more than in 2002 with 1650 students in fields such as education, health sciences, agronomy, social sciences, engineering, economics and administration, and mathematics and natural sciences.
3. According with the Expanded Science Citation Index, from 1990 to 2007 the number of scientific articles written by Colombian authors in indexed publications were multiplied by 5, which represent an annual average growth rate close to 10% (Dirección Nacional de Planeación, 2009).

4. In Colombia there was a total of 316 registered journals; 168 of them were registered in Directory of Open Access Journals (DOAJ) as Open Access Journals and were part of 30 repositories, according to the web ranking of repositories (Colciencias, 2010).

It can be noted that since the last two versions of CONPES seek access to information and the use of ICT, Open Access practices were strengthened, causing a considerable increase of scientific production in scientific journals and repositories. Other actions to be considered in the Open Access field were the measurement and classification of Colombian scientific journals, made by Colciencias, and the accreditation of academic programs and universities through the Ministry of Education, because these actions encouraged the creation of more magazines, as well as their coverage (Uribe-Tirado, 2016). Even though these measures were taken from the central government, the truth is that the efforts made to develop STI through Open Access strategies are led by educational institutions which are mostly private.

In order to issue the last CONPES 3582, in the year 2009, some constraints were identified which did not allow optimal results in SNT: low levels of innovation in companies, lack of human resources to do research, low appropriation of knowledge and low capacity to generate knowledge (Dirección Nacional de Planeación, 2009). Despite the progress made since the first CONPES with regard to the issues of publications and repositories, the efforts and strategies are still not enough, and that is the reason why, following the CONPES 3080 guidelines, this new document seeks to improve the current figures and the positioning of the country as a generator of development (Plata & Cabrera, 2011).

Related to intellectual property, and in order to keep up with the concepts in the document, knowledge cannot be fully free because of the fact that there is no appropriate estimation of the results of investment in knowledge generation, which results in insufficient incentives to invest and also leads to market failure. Consequently, this CONPES proposed a strategy to seek to improve the indicators in this subject, which goes against the existing actions in Open Access.

The strategy is divided into several actions; one of these actions is rewarding the efforts of innovation obtained from the use of intellectual property, and in turn it enhances knowledge transfer. It also seeks to strengthen the protection and use of research results by financing the cost of these registration processes, as well as the training of researchers to manage licenses and patents (Dirección Nacional de Planeación, 2009).

Concerning the transfer and democratization of information, and looking for more productive processes, it is suggested to maximize the use of ICT. In this sense, the promotion and expansion of the National Network of Research and Education are presented as a means to guarantee more coverage and participation of the actors who belong to the National System of Technology and Innovation (Dirección Nacional de Planeación, 2009).

Particularly regarding social appropriation of knowledge, the strategy proposes to spread the processes of research and innovation throughout every possible communication channel, but it does not include the use of new technologies of information (Dirección Nacional de Planeación, 2009).

At this point, the results of the analysis show that all policies—including this one—are aimed to protect data and research outcomes through licenses and patents, which would restrict their access and use, and of course would go against the measures taken in access to scientific publications (Golden Road) and repositories (Green Road).

The previous dichotomy is also observed in the National System of Open Access in Colombia, which promotes visibility, access, use and appropriation of knowledge produced in Colombia through Open Access (Sistema Nacional de Acceso Abierto de Colombia, n.d.), but it requires the use of licenses and different permissions other than those suggested by CONPES in STI in order to make visible innovation and production data.

In conclusion, although STI's latest public policies focus on sharing information and the use of ICT, these policies have not been sufficient to make possible the goals toward economic growth. Generally speaking, a central aspect which should be addressed is the use of Open Access in order to positively impact the economic and industrial growth of the country.

In this sense, Fig. 5.1 shows the evolution of Colombian GDP from 1990 to 2015, which has been increasing during the last few decades, as well as the percentage of R&D investments of the country, which also show a positive behavior but follow an unstable trend. This could suggest the need to increase the scale of the impact of the GDP changes over the STI development.

This necessarily implies the development of a specific policy of Open Access, its integration with the STI policy framework, and the establishment of the best strategy in order to take advantage of the link between the use of open access and the expected economic results.

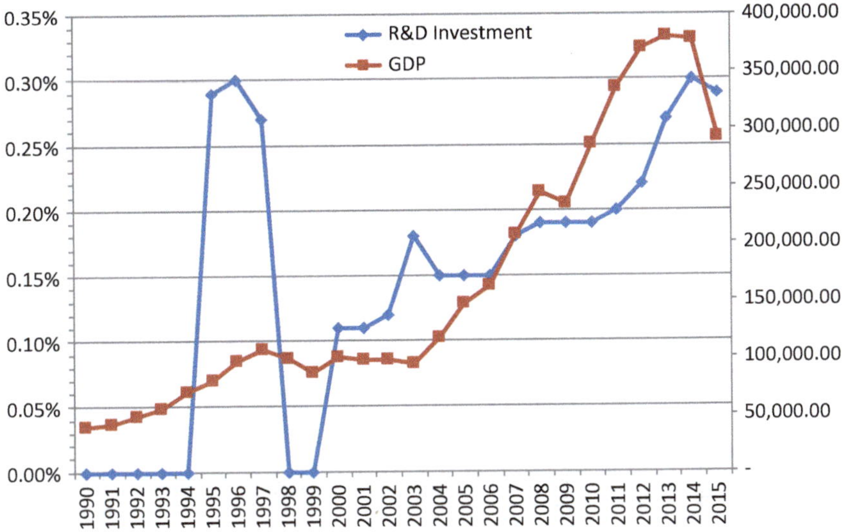

Fig. 5.1 Evolution of GDP (US$ millions) vs. R&D Investment (as a GDP Percentage) in Colombia, 1990–2015

For example, STI's policies could go jointly with Open Access policies and make creative commons licenses mandatory for every result of investigation that has been financed by the state. Therefore, not only the protection of the information would be guaranteed but also the responsible use of generated data.

Another important focus of the Open Access policy should be related to the standardization of the process in order to measure the results of Open Access as well as the research outputs, taking into account that currently there is not enough information to evaluate the relationship between R&D investment and the generation of Open Access results. Figs. 5.2 and 5.3 shows the evolution of Colombian journals with OA policies as well as the number of Open Access publications in Scopus, one of the most important scientific indexes. These results highlight the relevance of the Open Access movements at the global level, but at least in the case of Colombia, the gathered data is not sufficiently complete to establish the real evolution and the relationship between the R&D investments and STI results.

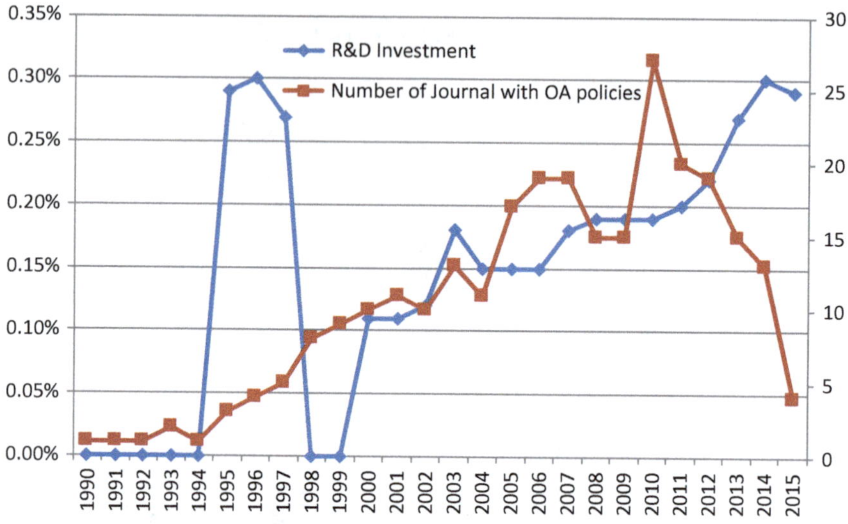

Fig. 5.2 Evolution of GDP (US$ millions) vs. R&D investment (as a GDP percentage) in Colombia, 1990–2015

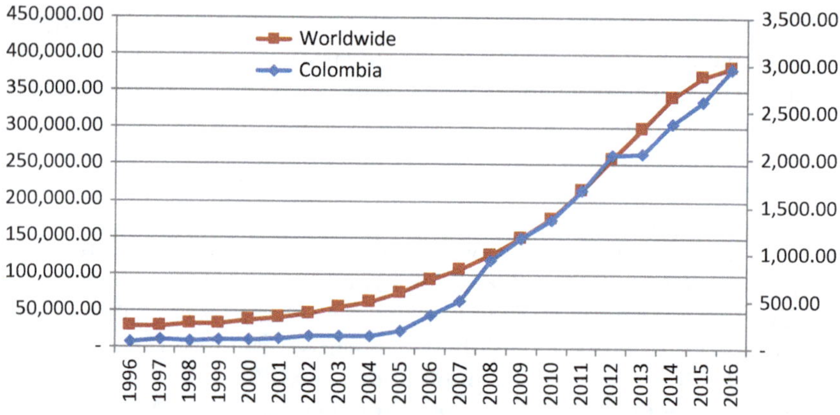

Fig. 5.3 Number of Open Access publications in Scopus. Colombia vs. Worldwide 1996–2016

According to Uribe-Tirado (2016), the number of Colombian scientific journals has experienced a significant increase in terms of Open Access that reaches approximately 95% registered in DOAJ, despite the lack of alignment between the expected results of scientific production and the use of open platforms. This misalignment originated because the incentives generated in academic institutions—which are the main producers of scientific information in the case of Colombia—are related to publications through indexes such as Web of Science and Scopus. These indexes have only about 20% of Open Access content and also publish almost entirely English content, which makes it difficult for non-academic actors to access scientific products.

In this sense, new impact measures should be developed to contrast the effectiveness and appropriateness of scientific contributions to solve social and economic problems in the national context in accordance with the characteristics of the access provided to the academic and non-academic actors.

5.5 Conclusions

After the analysis of different international experiences in the operationalization of Open Access in Science, one of the main conclusions is the heterogeneous nature of the dissemination of this topic, as well as the recent emergence of measurement of the impact on the STI performance.

Although, from a political and economic point of view, a relation between STI's policies and the economic growth of a country is accepted, some countries are aiming for the liberalization of information—via the golden and green routes—in order to promote science and technology. The result of this analysis could reveal a lack of understanding of the impact of Open Access and also a time lag between STI policies proposed in other countries and Colombia.

In spite of this, there are some policy implications that can be used in order to regulate the exploitation of the advantages of Open Science in order to enhance the performance and outputs derived from scientific knowledge. According to the analysis carried out, in this study we identify four key areas in order to address the implementation of Open Science in an emerging economy with the absence of regulation in this field, as in the case of Colombia.

These four areas are Integration with STI, Institutional Effort, Impact Measuring and Society Promotion. Table 5.2 compiles the policy strategies proposed as well as the suggested impacts on the STI performance. The presented implications are especially useful in contexts like Colombia, where the STI system has not incorporated specific policies to articulate

Table 5.2 Policy strategies to implement in countries with emerging economies (Colombia)

Policy Area	*Objective*	*Reference Country*	*Possible Impact on STI Outputs*
Integration with STI	To regulate particularly open access in science through the development of a law closely integrated with the STI CONPES documents.	European Union	To address the use of sources of science and technology in a more efficient way, avoiding duplication of efforts, effective design of incentives and improving the quality of outputs
Institutional effort	To generate a collaborative approach which involves the integration of all the actors, around the development of, access to, dissemination of and use of scientific outputs.	European Union/ Argentina	To guarantee the alignment of all actors involved in the STI process: Private, public, academic and non-academic.
Impact measurement	To drive the process of design mechanisms in order to gather and analyze data on the actual impact of open access in Colombia's science system.	European Union/ Australia/OECD	To control and stimulate the impact of the policies designed to promote the enhancement of STI performance
Society promotion	To promote a culture about Open Science in different spheres of society based on the availability of infrastructure and dissemination mechanisms.	European Union/ UK	To encourage the use of scientific outputs for the development of a more innovative society

the performance of Open Access and Open Science and therefore lacks measurement instruments for monitoring the results and impact of these mechanisms on the improvement of innovation results.

These strategies can be materialized with possible "mandates" in Open Access. Some of them are:

1. To make available the information of the research projects with governmental resources. This includes self-archiving (Green Road) using licenses like creative commons to make accessible and exploit (re-use) the information.
2. To adopt the figure of "embargo" as an exception and it could not be extended for an unlimited time. Some countries establish that the publication must be done, to consult online and without cost, no later than one year after the research has been completed.
3. To implement a standard that measures the quality of the publications and the impact of the public investment on the shared information that has been established as a priority in STI.
4. To instruct researchers about the operation of institutional repositories and self-archiving of materials protected by copyrights.
5. To align the incentives utilized for the academic and research institutions with the expected outputs in terms of STI, in order to generate research with high quality and visibility standards available through open platforms to increase the use for non-academic actors as well.

Funding

The second author acknowledges the financial support of Colciencias through the project "Dinámicas de Innovación en Contextos Poco Intensivos en I+D: el caso de la Industria Manufacturera Colombiana" grant 121577657885, as well as the financial support of the Universidad del Norte through grant 2017–18.

References

Australian Academy of Science. (2014). *The Importance of Advanced Physical and Mathematical Sciences to the Australian Economy*. Retrieved from https://www.science.org.au/supporting-science/science-sector-analysis/reports-and-publications/importance-advanced-physical-and

Australian Government. (2017). *Australian National Science Statement*. Retrieved from http://science.gov.au/scienceGov/NationalScienceStatement/index.html

Australian National Data Services. (n.d.). *About Us*. Retrieved from https://www.ands.org.au/about-us

Beagrie N., & Houghton J. (2014). *The Value and Impact of Data Sharing and Curation: A Synthesis of Three Recent Studies of UK Research Data Centres*. JISC, Bristol. Retrieved from http://commons.esipfed.org/sites/default/files/JISC%20report%2C%20The%20Value%20and%20Impact%20of%20Data%20Sharing%20and%20Curation%2C%202014.pdf

Budapest Open Access Initiative. (2002). Retrieved from http://www.budapestopenaccessinitiative.org/read

Cabrera, K. I. (2014). Modelos de acceso abierto en educación y ciencia. *Educación y Educadores, 17*, 321–338.

Cabrera, K. I. (2015). Análisis comparativo de políticas públicas en modelos de acceso abierto en América Latina. El caso de Brasil y Argentina. *RUSC. Universities and Knowledge Society Journal, 12*, 15–25.

Chan, L., & Costa, S. (2005). Participation in the Global Knowledge Commons: Challenges and Opportunities for Research Dissemination in Developing Countries. *New Library World, 106*, 141–163.

Colciencias. (2010). *Balance de gestión Colciencias 2006–2010*. Retrieved from https://colaboracion.dnp.gov.co/CDT/Desarrollo%20Empresarial/Informe_%202006-2010.pdf

Costa, M., & Leite, F. (2016). Open Access in the World and Latin America: A Review Since the Budapest Open Access Initiative. *TransInformação, 28*, 33–45.

Dirección Nacional de Planeación. (1980). *Plan de integración Nacional. Política de Ciencias y Tecnología*. Retrieved from https://colaboracion.dnp.gov.co/CDT/Conpes/Econ%C3%B3micos/1640.pdf

Dirección Nacional de Planeación. (1991). *CONPES 2540*. Retrieved from https://colaboracion.dnp.gov.co/CDT/Conpes/Econ%C3%B3micos/2540.pdf

Dirección Nacional de Planeación. (1994). *CONPES 2739*. Retrieved from https://colaboracion.dnp.gov.co/CDT/Conpes/Econ%C3%B3micos/2739.pdf

Dirección Nacional de Planeación. (2000). *CONPES 2739*. Retrieved from https://colaboracion.dnp.gov.co/CDT/Conpes/Econ%C3%B3micos/3080.pdf

Dirección Nacional de Planeación. (2009). *CONPES 3582*. Retrieved from https://colaboracion.dnp.gov.co/CDT/Conpes/Econ%C3%B3micos/3582.pdf

Dirección Nacional de Planeación. (n.d.). *El Consejo Nacional de Política Económica y Social, CONPES*. Retrieved from https://www.dnp.gov.co/CONPES/Paginas/conpes.aspx

European Commission. (2017a). *Open Science*. Retrieved from http://ec.europa.eu/research/openscience/index.cfm?pg=openaccess

European Commission. (2017b). *Open Science Monitor*. Retrieved from http://ec.europa.eu/research/openscience/index.cfm?pg=home&section=monitor

European Commission. (n.d.). *Data Management*. Retrieved from http://ec.europa.eu/research/participants/docs/h2020-funding-guide/cross-cutting-issues/open-access-data-management/data-management_en.htm

Houghton, J. (2009). *Open Access – What Are the Economic Benefits? A Comparison of the United Kingdom, Netherlands and Denmark*. Melbourne: Victoria University.

Houghton, J., & Sheehan, P. (2009). Estimating the Potential Impacts of Open Access to Research Findings. *Economic Analysis and Policy, 39*, 127–142.

Houghton, J., Rasmussen, B., & Sheehan, P. (2010). *Economic and Social Returns on Investment in Open Archiving Publicly Funded Research Outputs. Report to the Scholarly Publishing and Academic Resources Coalition (SPARC)*. Melbourne: Center for Strategic Economic Studies of Victoria University.

LAReferencia. (n.d.). *¿Quiénes somos?* Retrieved from http://www.lareferencia.info/joomla/es/institucional/quienes-somos

Lasthiotakis, H., Kretz, A., & Creso, S. (2015). Open Science Strategies in Research Policies: A Comparative Exploration of Canada, the US and the UK. *Policy Futures in Education, 8*, 968–989.

Levin, N., Leonelli, S., Weckowska, D., Castle, D., & Dupré, J. (2016). How Do Scientists Define Openness? Exploring the Relationship Between Open Science Policies and Research Practice. *Bulletin of Science, Technology & Society, 36*, 128–141.

Ministerio de Ciencia, Tecnología e Innovación Productiva. (n.d.). *Sistema Nacional de Repositorios Digitales*. Retrieved from http://repositorios.mincyt.gob.ar/index.php

OECD. (2015). *Making Open Science a Reality*. OECD Science, Technology and Industry Policy Papers No. 25, Paris: OECD Publishing. https://doi.org/10.1787/5jrs2f963zs1-en

OECD. (2017). *OECD Science, Technology and Industry Scoreboard 2017*. Paris: OECD Publishing. https://doi.org/10.1787/9789264268821-en

Organization of American States. (1972). *Meeting on Scientific-technological Priorities and Economic and Social Planning*. Retrieved from http://unctad.org/meetings/en/SessionalDocuments/CSTD_2013_DraftRes_STI.pdf

Picasso, V., & Phelan, L. (2014). La evolución del acceso abierto a la investigación y a los datos en la educación superior en Australia. *RUSC. Universities and Knowledge Society Journal, 11*, 128–141.

Plata, L. C., & Cabrera, K. I. (2011). La normativa colombiana sobre propiedad intelectual: un análisis de la política pública en ciencia, tecnología e innovación a partir del desarrollo económico. *Revista Opinión Jurídica, 20*, 87–104.

Presidencia de la República de Colombia. (1996). Colombia: Al filo de la Oportunidad, Colciencias, Colombia. Retrieved from http://www.plandecenal.edu.co/cms/media/herramientas/colombia_al_filo_de_la_oportunidad.pdf

RENATA. (n.d.). *¿Qué es RENATA?*. Retrieved from http://www.renata.edu.co/que-es-renata/

Research Council UK. (n.d.-a). *Aims and Organisation*. Retrieved from http://www.rcuk.ac.uk/about/aims-and-organisation/

Research Council UK. (n.d.-b). *RCUK Policy on Open Access and Supporting Guidance*. Retrieved from http://www.rcuk.ac.uk/documents/documents/rcukopenaccesspolicy-pdf/

RICYT. (2017). *El estado de la ciencia. Principales indicadores de Ciencia y Tecnología Iberoamericanos/ Interamericanos*. Buenos Aires: RICYT. Retrieved from http://www.ricyt.org/files/Estado%20de%20la%20Ciencia%202017/El_Estado_de_la_Ciencia_2017.pdf

Sistema Nacional de Acceso Abierto de Colombia. (n.d.). *¿Qué es el SNAAC?*. Retrieved from http://190.242.114.6:8080/web/guest/estrategia-snaac

UK Parliament. (n.d.). *The Finch Report: A U-turn in UK Open Access Policy*. Retrieved from https://publications.parliament.uk/pa/cm201314/cmselect/cmbis/99/9906.htm#a3

Uribe-Tirado, A. (2016). El Acceso Abierto en Colombia. Un camino por recorrer. *Revista Cuadernos Fronterizos, 12*, 33–36.

CHAPTER 6

The Inclusion of Ecuadorian Women in Scientific-Technological Fields

Antonio A. Franco-Crespo, Fernando Herrera García, and Jessica Lin Gutiérrez

6.1 Introduction

Human history has several moments with deep conflicts linked to the use of power by the strongest members of society when taking advantage of dominion over physical, economic or knowledge-related resources. These problems have evolved throughout time without cessation up until the present day. People have demonstrated a duality in his actions, which, on one hand, makes him a thinking being endowed with a creative intelligence allowing the development of art and science and, on the other hand, makes him an irrational force capable of cruelly destroying his own kind.

Science and technology have been fundamental in the development of our society (Catton, 1980). They are complex social constructs associated with human, social, cognitive and material factors (Cañedo Andalia, 2000), which use and produce knowledge. These complex fields have

A. A. Franco-Crespo (✉) • F. Herrera García • J. Lin Gutiérrez
Department of Organizational Studies and Human Development, Escuela Politécnica Nacional, Quito, Ecuador
e-mail: antonio.franco@epn.edu.ec; edwin.herrera@epn.edu.ec; jessica.lin@epn.edu.ec

C. I. Pardo Martínez et al. (eds.), *Analysis of Science, Technology, and Innovation in Emerging Economies*,
https://doi.org/10.1007/978-3-030-13578-2_6

not been without disputes and inequity situations, one of which is the systematic exclusion of half of the world population: women. Science and technology could be cause or solution of conflicts regarding equity when men use their power on women.

It is true that the scientific and technological contributions made by women are proportionally smaller than those of men (Guardia, 2013). In spite of this, their contributions are not less relevant, especially when said that these contributions have been accompanied by the historical struggle in accessing academic and educational contexts. In Ecuador, the inclusion of women in the academic, scientific and technological fields has been attached to many difficulties. These problems are similar to those found in other countries in the region.

In recent years, a more equitable environment has emerged due to the formulation of new policies and reforms in the Constitution of Ecuador. Articles 11, 66 and 331 of the Ecuadorian Constitution reaffirm the stance taken against discrimination and in favor of equality (Asamblea Nacional del Ecuador, 2008). The Constitution, along with other societal norms, strengthens the irrevocable rights of all members of society.

Ecuador demarcated 12 objectives in its National Plan for Good living in 2009. It explicitly defined that "The State recognizes and guarantees the indivisible character of the system of rights of good living and, above all else, ensures its institutional and public realization. Therefore, this objective proposes the application of integrated policies, capable of tackling the complexity of exclusion and promoting new senses of cohesion and (re)distribution within the framework of recognizing diversity" (Consejo Nacional de Planificación, 2009).

Ten years after the adoption of the new Constitution of Ecuador and the implementation of diverse legal norms in the fields of science and technology, the objective of this investigation is to understand the role of women in Ecuador within these fields. Several important advances are recognized by the actors involved in the scientific and technological development in Ecuador. However, eliminating the gap in women's participation is still a pending goal.

This chapter will focus on the understanding of gender inclusiveness in the Ecuadorian context of scientific and technological fields, beginning with the modification of the legal framework during the period from 2007 to 2017.

6.2 Literature Review

The dispute regarding conceptions about the role of women has become permanent. The debate has been focused on subjects that have evolved and regressed throughout the passage of the centuries and decades. The participation of women in education, in the workforce in the home or in motherhood has not been defined on part of women themselves. They have been imposed by the state or by men, placing them in stereotypical roles of single, married or working women (Dalla & James, 1977).

History questions and answers depend of cultural conceptions with characteristics that aren't necessarily temporary, but rather have permanent heuristic elements (Sánchez, 2005). The evolution of the role of women within society is marked by discrimination and subordination, suffered at the hands of a society whose culture has been strongly influenced by men (Guardia, 2013). Since the prehistoric period, women have formed an extremely important role within society, occasionally being considered goddesses in some parts of the world, and becoming significant political figures in others, such as Cleopatra of Egypt, who, despite her political position, has been considered inferior compared to other leaders (Mora, 2013).

One of the most fundamental rights is education, which was globally recognized in the year 1948 when it was guaranteed in the Universal Declaration of Human Rights (United Nations, 2015). Without a doubt, education is the driving force behind integrated development of human beings in society. It is that which makes one an active member of society by including both individual and collective scopes (Blanco, 2005). However, despite the importance of education for both men and women, access to education for women has only recently emerged in history, a product of an exclusive governmental and social system which denies and impedes women from accessing education.

In the nineteenth century in Latin America, women received a rather conservative education, based on the teaching of house hold chores like cooking or sewing, with the objective of educating perfect mothers and wives that would be able to manage a home and their families (Wright de Kleinhans & Alvarado, 2005). Afterward, education was designed differently according to the financial situation of a family, focusing on the fact that women belonging to the upper class were to receive a more academic education, while women of lower classes were trained to work in manual labor (Wright de Kleinhans & Alvarado, 2005).

In general terms, the female presence is beginning to become more and more evident. Women began to leave traditional roles, rejecting the oppression of which they were victims in a patriarchal system, which has allowed the contributions of many women who fought for privileges that were exclusive to men to go unnoticed (Aquino & Támez, 1998). One of these women was Sor Juana Inés Cruz, who is considered one of the first feminists of the New World by standing out in a world where knowledge belonged to men and where men were the only ones with the right to access university level education. Cruz confirmed that one could be female and intelligent at the same time (Dufort, 2011).

In this period in Ecuador, the integration of women into social life was even more exclusive toward indigenous and black women by imposing racist and classist ideologies which prevented them from accessing any kind of connection with formal education (Aquino & Támez, 1998). In this unfair colonial society, the lack of access to academic knowledge did not impede the apparition of people such as Manuela Sáenz, who were doers of the impossible and did not accept social marginalization and political imposition; they headed the process of emancipation for both genders.

During the republican era in Ecuador, which began in 1830, one can identify the different points in time that standout from the consolidation of the patriarchal model or of its breakdown. One of the most notable points was during the 1860s and 1870s, when Gabriel García Moreno became a significant stakeholder as president of the nation by strengthening the influence of the Catholic Church in the education of women under the idea of education as a focal point of the family, born to be exemplary mothers and wives (Maiguashca, 1994). Education in the hands of the Catholic church (which was patriarchal in ideology) under the government of García Moreno allowed a limited group of women to access primary education by permitting the establishment of two schools for girls of the already existing 50 (Maiguashca, 1994).

The predominant ideology in the middle of the nineteenth century and the one that would continue for many decades in Ecuador converted education into a right that was practically unreachable for women of the time. People considered education to be important for women, seeing as how it was women that influenced the education of men, who would be tasked in the future to formulate, implement and govern their families, organizations and the state, reinforcing discrimination (Goetschel, 1999).

Starting in 1900, education in Ecuador experienced an extensive shift, becoming a fundamental pillar of Ecuadorian society by being considered it a political and governmental priority. This change was made mainly for the female population by providing spaces in which they could develop themselves and gain both independence and autonomy (Ayala Mora, 1996). The new liberal vision, led by Eloy Alfaro, allowed the twentieth century to begin with a secular education system, independent of the Catholic Church's view of women, family and society. There was an incentive toward development of abilities that prepared women to fulfill a more active role (Ayala Mora, 1996).

The political and economic changes of that point in time provoked the need for new administrative, educational and manufacturing jobs that were given to women (Goetschel, 1999). The teachers of the time played an important role in the social changes of the following decades, driving the composition of syndicates and student and teacher unions that were always at the front of action with respect to educational reform that guaranteed the quality thereof (Goetschel, 1999).

However, in Ecuador, the access to higher education centers was a right reserved exclusively for men, a practice which continued until the republican era when Matilde Hidalgo graduated with a bachelor's in medicine in 1919, becoming the first women to attend university and obtain a university degree (Estrada, 2016). The path in achieving this was filled with difficulties; there were multiple rejections and obstacles that needed to be overcome in order to finish her university studies. Obtaining the right to vote was equally complex; this occurred in 1924, after a collective militant effort on part of Matilde Hidalgo and other notable women (Estrada, 2016).

The period occupied by higher education in Ecuador until the year 2006 was accentuated by a deficient public policy that presented diverse problems, from the limited access to university education and the poor quality thereof, to the weak management of new scientific and technological techniques (Charvet, 2016). Since this year, extensive changes have emerged in Ecuadorian higher education, starting with Rafael Correa's victory in the presidential election along with his political movement, "Alianza Pais", in November of 2006. Correa, during his electoral campaign, presented a government plan that would reclaim old social ambitions like free, quality and significant higher education (Alianza Pais, 2006), which showed a vision for the Ecuadorian educational system in the event that he was to win the election.

The tipping point in the administration of the Ecuadorian educational system can be recognized as the approval of the 2008 Constitution of the Republic. This supreme legal normal explicitly declares Ecuador as a state that recognizes education as a permanent right throughout the lives of the people, focused on human beings within the framework of respect for human rights, a sustainable environment and democracy. Within this framework, education would be participative, obligatory, intercultural, democratic, inclusive, diverse, quality and friendly; it would encourage gender equality, justice, solidarity and peace (Asamblea Constituyente del Ecuador, 2008). Additionally, the constitution requires that the state guarantees equality and financial and social inclusion regarding its people, seeking to avoid any kind of discrimination (Asamblea Constituyente del Ecuador, 2008).

The National Plan for Good Living between 2009 and 2013 explicitly established, for the first time in Ecuador, a new model for accumulation and distribution, which changed from the system of the time to another based on diversification and replacement of exports, outcomes of research, innovation, science and technology, all tied to a primary export model.

The National Plan for Good Living (2009–2013) explicitly established, for the first time in Ecuador, a new model for accumulation and distribution, which changed from the system of the time which was tied to a primary export model, to another based on diversification and replacement of exports that are the fruit of research, innovation, science and technology, being the highest stage of an economy based in bioknowledge and tourist services (SENPLADES, 2009).

The definition of placing science and technology as fundamental pillars to invigorate society from distinct social, economic and political perspectives is not new to the world (Jara, 2014), although it is a complete novelty for Ecuador (SENPLADES, 2009). In every case, a coincidence exists that education is what places foundation for growth and evolution with respect to the processes of formulating public policies that lead to the completion of a development model based on information, science, technology and innovation (Jara, 2014).

In the year 2010, the *Ley Orgánica de Educación Superior* (LOES "The Organic Law of Higher Education") was approved. This radically changed higher education in the country. This law determined that undergraduate higher education be free. It also created the *Consejo de Evaluación,*

Acreditación y Aseguramiento de la Calidad Superior (CEAACES "The Committee of Evaluation, Accreditation and Higher Quality Assurance"). This law also demanded the investment of a minimum of 6% of the university budget to be dedicated research as well as the regulation of the minimum amount of teaching fellows and professors with doctorates in the universities and technology institutes (IES "Institutions of Higher Education"). The law also covered other important aspects of higher education (Asamblea Nacional del Ecuador, 2010).

At the same time that LOES, its regulations and norms were being implemented, the institution of higher education process of evaluations was also proceeding in order to fulfill the twentieth transitory provision of the Constitution of the Republic that stipulates: "In the period starting from the date of validity of this Constitution, all institutions of superior education, as well as majors, programs and graduate degrees, must be evaluated and accredited in conformity with the law. In the event that the evaluation and accreditation is not passed, it will not form part of the system of higher education" (Asamblea Constituyente del Ecuador, 2008).

The process of evaluation of Ecuadorian universities required the institutions of higher education to redirect their institutional practices, which did not stop 14 out of 71 universities from being closed in 2012 on part of the governing body of university evaluation in Ecuador, the CEAACES. These institutions did not demonstrate quality, relevance and character for the general good of education (Long, 2013). The final evaluation reports of the CEAACES detected alarming shortcomings, such as the total absence of full-time professors in a few of the institutions evaluated. In others, there were zero instructors with PhDs, and a complete absence of research and publications (Long, 2013).

Various indicators could demonstrate the transformation of Ecuadorian higher education and, therefore, the contribution of these indicators to the production of knowledge in the last decade. Tables 6.1 and 6.2 show the data of the number of publications made by Ecuador and its neighboring countries in 2007 and 2016 (SCImago, 2018).

In the period since 2007 to 2016, the number of documents published in Ecuador increased considerably, from 345 documents in 2007 to 2249 in 2016 (SCImago, 2018). This is an important achievement in relation to the countries that lead the Latin American rankings, such as Brazil and Colombia. In fact, Ecuador multiplied by six the number of publications in that period.

Table 6.1 The data of the number of publications made by Ecuador and its neighboring countries (2007–2016)

Rank	*Country*	*Documents*	*Citations per document*	*H index*
16	Bolivia	229	24.15	103
12	Ecuador	345	20.2	126
9	Peru	701	26.78	179
6	Venezuela	2019	11.42	181
5	Colombia	2510	15.72	216
4	Chile	5293	18.31	291
3	Argentina	8022	18.56	337
1	Brazil	36,573	14.82	461

Source: SCImago, 2018

Table 6.2 The data of the number of publications made by Ecuador and its neighboring countries (2007–2016)

Rank	*Country*	*Documents*	*Citations per document*	*H index*
16	Bolivia	312	1.22	103
9	Venezuela	1476	1.06	181
7	Peru	2228	1.13	179
6	Ecuador	2249	0.85	126
5	Colombia	9481	0.79	216
4	Chile	12,448	1.12	291
3	Argentina	12,864	0.94	337
1	Brazil	68,908	0.81	461

Source: SCImago, 2018

The progress and the constant struggle to improve the education system in Ecuador have had good results. However, the existing information does not allow an adequate understanding of the contribution of women, and the impact on them.

In Ecuador, the new focus of governmental strategies toward an economy is based on knowledge. This is in tandem with transformations in the higher education system that have been caused by processes of evaluation and the implementation of state regulations. These processes were not necessarily accompanied by improvements in gender equality in these fields despite the demand made through the obligatory evaluation models for all institutions in the university system (CEAACES, 2014).

The number of women enrolled in institutions of higher education has surpassed that of men, the 55% of university students are women (Senescyt, 2015). However, women still tend to study majors related to roles that are traditionally considered to be for women, such as service and caring for a family (Senescyt, 2015). In fact, while the percentage of women in fields like education, family, health or administration is at 63%, the percentage of women in engineering, physical, mathematical and agricultural sciences barely reaches 28% (Ponce, 2016).

Female professors at higher education institutions, in the year 2012, composed a minority group compared to men. Just around 43% of higher education institutions scored 1.0/1.0 in the gender indicator (CEAACES, 2014). For the year 2015, 12 out of 13 evaluated universities reported having between 30% and 70% of managerial positions at the institution of higher education filled by women; seven of them had a percentage of female instructors higher than 40% (CEAACES, 2016). On the other hand, the participation of women in the field of research in universities demonstrates a great disparity; barely 4% of university instructors perform research (Senescyt, 2015).

Table 6.3 shows the percentage of women's participation in the academic field, whose data was extracted from the evaluations made to universities in 2015.

Table 6.4 shows the number of women who have traveled abroad for studies increased significantly between 2011 and 2016. While in 2011, 1448 women traveled, in 2014, 5188 did so, increasing to 8760 in 2016 (Instituto Nacional de Estadísticas y Censos, 2016). In just two years, the number of women who were dedicated to research in Ecuador, increased

Table 6.3 The percentage of women's participation in the academic field

Women's participation	*Female*
Managerial positions at the institution of higher education	30%–70%
Female instructors in higher education	40%
Women in the field of research in universities	4%
Female university students	55%
Female students in engineering, physical, mathematical and agricultural sciences	28%
Female students in fields like education, family, health or administration	63%

Source: CEAACES, 2016

Table 6.4 The number of women who have traveled abroad for studies

	2014	*2016*
Women who have traveled abroad for studies	5188	8670
Women who are dedicated to research	3065	4688

Source: Instituto Nacional de Estadísticas y Censos, 2016

from 3065 to 4688, between 2014 and 2016 (Instituto Nacional de Estadísticas y Censos, 2016).

This disparity can also be found with respect to access to scholarships linked to fields about life sciences, production and innovation, where there is a minority of women in those fields compared to the number of women in other fields of study (Ponce, 2016).

Equality within the university education system is not a condition that is exclusively determined by constitutional access to higher education, but rather also by the elimination of any kind of harmful discrimination that is found to be directed toward students, instructors, researchers, coordinators or administrative personnel (Charvet, 2016). Due to this, various public actions have been defined in the face of evidence of problems regarding equality for women in Ecuador. The different versions of the *Plan Nacional del Buen Vivir* ("National Plan for Good Living") have clearly established the government's intention to improve conditions for vulnerable groups. The National Plan 2013–2017 provides 12 objectives that call for "a society in which people can satisfy their needs and achieve a dignified life and death free of a kind of violence and discrimination, with equality and social justice, achieving harmony with oneself, society and nature" (Consejo Nacional de Planificación, 2013).

Additionally, there are other government efforts focused on strengthening the right to equality, mainly with respect to gender, considering that this principle "demands that the differences in needs, interests, behaviors and aspirations of men and women with respect to the same and equal rights be considered, valued and promoted in the same way" (García, 2008). One of the strategies of public policy implemented in this context was the formulation of the *Agenda Nacional de las Mujeres y la Igualdad de Género* ("The National Agenda of Women and Gender Equality"), which is substantiated by 12 pillars, one of which is education and knowledge. This references the social transformation by means of the production of knowledge and the existence of an integrated educational system that has overcome gender discrimination (Consejo Nacional de Igualdad de Género, 2014).

The Agenda considers the intention of strengthening and consolidating the integrated transformation of the educational system and production of knowledge as one of its policies in order to overcome patterns of gender discrimination and create conditions, content and practices of full-fledged equality, harmonious coexistence and reciprocation between all people (Consejo Nacional de Igualdad de Género, 2014). In order to achieve this, various strategies have been formulated, one of which lays out the obligatory nature of "incorporating affirmative action mechanisms in the processes of scientific, academic and technological innovation in the country in order to benefit from the talents of women and LGBTI individuals" (Consejo Nacional de Igualdad de Género, 2014).

In this context of change toward a new model of development in Ecuador, where knowledge is the basis of its attainment, the role of women has been favored through the formulation of explicit norms in defense of gender equality. However, the efficacy of these declarations for true equality for women within the scientific and technological fields in Ecuador has yet to be determined.

6.3 Methods

The present research about gender inclusiveness and public policies in the period 2007–2017 is related to the intent to transform the Ecuadorian reality. It was oriented toward action, social transformation and emancipation. Because of this, this research is framed within the paradigm of critical theory (Guba & Lincoln, 2000).

Due to the nature of the intent of this research and its focus on the explanation of the reality in Ecuador with respect to academia, public policy and gender relations, the qualitative focus was a method selected because of its coherency with the paradigm of the chosen research. People are considered as a holistic element, not as variables, considering the context, history and the current situation (Creswell, 2007). In accordance with what is stipulated in the theory (Quecedo Lecanda & Casatño Garrido, 2003), the design of the qualitative research was flexible and could be adjusted to sources of information found throughout the research.

Qualitative research uses means of gathering information that allows a wide understanding regarding the context of the subject of the study using documentary analysis, observation, interviews or analysis of the subject themselves (Creswell, 2007).

The type of qualitative research selected was a case study, understood as a kind of research that allows one to obtain internal knowledge regarding social life, which describes and interprets social phenomena from the perspective of the social, holistic and natural environment and of the character involved (Murillo & Martinez-Garrido, 2010). Throughout the case study, a narrative description was produced about the subject of the study, which was the *Red Ecuatoriana de Mujeres Científicas* (REMCI "Ecuadorian Network of Female Scientists").

The sources of information for this investigation were (1) the main legal norms related to the public policies regarding science, technology and gender; (2) statistical data about the operating state bodies dealing with science; and (3) semi-structured interviews with women that work in the scientific and technological fields in Ecuador.

The documental analysis was limited to the period between the years 2007 and 2017, during which Rafael Correa was the president of Ecuador. The statistical information was mainly provided by the Ecuadorian Secretary of Higher Education, Science Technology and Innovation (Senescyt), and the Ecuadorian Committee of Evaluation, Accreditation and Quality Assurance of Higher Education (CEAACES).

Ten extensive interviews with members of REMCI were conducted, using a guideline made for the study. Both in person and online interviews took place, in the event that the researchers were not in the city of Quito at the time. Saturation criteria were used in order to restrict the number of interviews conducted (Creswell, 2007). The rise in information during the interviews allowed an adequate understanding of the interviewees' points of view, thereby facilitating codification of the interviews. Afterward, an analysis of the codes was performed, and they were grouped in order to define different categories, following Saldaña's proposal (2009).

In order to process interviews, codify and categorize said codifications of the interviews, the software NVivo v.11 was used, the results of which can be seen in the following section. NVivo is a qualitative data analysis computer software which offers a set of tools for working with text and multimedia information, where deep levels of analysis are possible.

6.4 Results

The findings of the study revealed a few of the challenges and problems which women, who have ventured into the fields of science and technology in search of gender equality and decent conditions for coexistence, deal with. Unfortunately, this is a milestone that all societies have pursued throughout history; it has not been an innate right. The evidence found reveals the existence of stereotypes and behaviors that still prevail in Ecuadorian society despite the implementation of public policy explicitly related to gender inclusion in the last few decades.

The analysis of the interviews conducted for the study included transcription and codification thereof. Said codification was done in a software called NVivo 11.0, in which the codes have node denominations. There are 22 recurring codes which can be seen in Table 6.5. The codes represent the central ideas and concepts identified in the interviews, whose frequency, in all the information, is given by the number of references.

Table 6.5 Recurring codes

Code	*Resources*	*References*
Positive change	7	14
Professional support	3	3
Understanding of gender discrimination	3	4
Visibility	3	4
Empowerment	1	1
Hierarchical positions	6	11
Family	6	11
Limited participation	2	2
Existence of stereotypes	8	14
Antiquated submissive roles	3	3
Insufficient change	2	2
Important part of society	1	1
Gaps	3	3
Wage gaps	3	3
Discrimination	6	12
Many roles	5	6
Workplace harassment	4	5
Lack of recognition	4	5
Deficient policy execution	9	13
Foreign scholarships	4	4
Public policies	10	14
Foreign studies	5	5

Table 6.6 Categories

Category	*Code*
Positive change	Positive change Professional support Understanding gender discrimination Visibility Empowerment
Work/family roles	Hierarchical positions Family
Problems that women face in the fields of science and technology	Limited participation Existence of stereotypes Antiquated submissive roles Insufficient changes Important part of society Gaps Wage gaps Discrimination Many roles Workplace harassment Lack of recognition
Public policies	Deficient execution of policies Foreign scholarships Public policies Foreign studies

On the other hand, the resources column shows the number of resources where the same node repeats itself.

Using the nodes that are shown in Table 6.5, the information was classified into categories that can be found in Table 6.6. The categories, according to theory, constitute the main findings of the research (Saldaña, 2009). Therefore, these categories represent the topics that have been discovered and that should be discussed in depth.

The categories found are related to both identified positive changes in the period of analysis and the prevailing problems that women face in the fields of science and technology.

The research found that there have been important advances in Ecuador related to scientific and technological production, as well as in the formulation and implementation of public policies that aim to reduce inequality for women. However, the perception of female researches is that the gap continues to be too large to be considered an advance. In fact, in Ecuador, "norms are [occasionally] written, but they are never materialized or put into practice" (Entrevista 1, 2017).

The list of problems and barriers that the interviewed women recognized with respect to their inclusion in fields of science and technology is quite long and includes limited female participation, presence of gender stereotypes and the definition of submissive roles for women. What is more, the existence in wage gaps, sexual harassment and lack of recognition also form part of these problems and barriers.

In seems to be that the problems that women face in the fields of science and technology are very similar to those found in other areas of society. These problems are related to the masculine anthropocentric models of the concepts of family, work, education, religion and the state, which are all predominant in society (Artigas & Turbine, 2007).

Discrimination toward women can have very subtle but profound tones, which ends up rooting itself into the culture of societies, which then converts it into "a cultural issue that has rooted itself into the people's way of thinking" (Entrevista 5, 2018). Everyday language replicates these subtle levels: deeming a female research as "doctorita" (a diminutive form of the word for female doctor) (Entrevista 2, 2017) or asking her to be a "secretary" (Entrevista 2, 2017). These are examples of this phenomenon.

The existence of gender stereotypes in Ecuadorian culture has occasionally been adopted by women as well, appropriating this rhetoric instead of discussing it. These recurrent roles and stereotypes in work and family spaces are known but not confronted. In the words of one of the interviewees:

> In the Name of the Daughter, a Tania Hermida movie that I love, there are a few children that make a comparison between the liberal and conservative era in Cuenca. There is a dialog between the children in which one of the boys says: 'You can't be the dad', and a girl says 'Who says I can't be the dad? I want to play and I want to be the dad.' The boy says 'No Manuela, it's because you're a girl,' to which she responds, 'Says who?' The little boy says 'Ay! Manuela no one said so but everyone knows.' I love this sentence because it is a way of maintaining symbolic order of reality…to say that no one said it, but that everyone knows. (Entrevista 2, 2017)

From all the findings of the research, it has been established that women in Ecuador do not benefit from the same opportunities in order to advance professional and personally within the fields of science and technology.

The existence of employment and wage gaps, as well as workplace harassment, hinders adequate professional development for female researchers. Lack of recognition reinforces exclusion practices, allowing that "the majority of research work continue to be given to men" (Entrevista 4, 2017). The work environment to which women, not only female scientists, are exposed is complex, and, "despite the fact that an attempt has been made to reduce levels of harassment, it is a reality that is present" (Entrevista 1, 2017). In terms of wage gaps, it is striking that even "until nowadays, Ecuadorian women earn 70% of when men earn. Clearly, this is not the case in every situation, but it is generally as such" (Entrevista 1, 2017).

Additionally, within the areas of work in which female scientists do advance, there is an important and significant gap with respect to access to positions of authority or strategically important positions. The majority of the time, the work of women has been lessened, and, despite fulfilling all the requirements needed to fill these positions, they are unable due to the lack of support and recognition in the work that they do.

Despite all the evident problems, research shows that there have been changes and advances in defense of women's rights. Thanks to the Ecuadorian Network of Female Scientists, women have been empowered in the fields of science and technology. The search for equality should definitely be a task for everyone involved, and not just at the government level. By means of education, it is possible to expect better days for female researchers, sensitizing students and instructors in the university system to topics regarding gender equality. It should be as such that they question "androcentrism in science and history" (Entrevista 7, 2017).

A woman who has made her way into fields of science and technology occasionally finds contradictions between her work and family life. Maternity and raising children can be a source of discrimination and not of wellbeing for female researchers. Occasionally, employers do not consider the role of women within their families, and place demands that do not take into account the rights of women as mothers (Entrevista 7, 2017).

Ecuadorian society has the necessary legal norms in order to ensure coexistence according to respect for both women and men, avoiding the elicitation of harmful situations for everyone, which is a product of gender discrimination. As mentioned by the interviewees, laws are important, and people are clearly working under norms and regulations. However, in this society, the cultural component creates obstacles and barriers for the effective application of said norms and regulations. The existence of laws does not guarantee in and of itself the complete fulfillment thereof.

The government is obligated to search for mechanisms that help the correct execution of said laws and to avoid misconduct of those who live under its jurisdiction.

According to the perception of women within scientific fields, one of the actions taken by the government which helped to reduce the gap in some way is access to scholarships in order to carry out graduate programs abroad. This allowed many women to leave the country to fulfill their professional goals without neglecting the nuclear family, seeing as how many of these scholarships given out foresaw the possibility of bringing their families with them.

Several strategies emerge from the research findings. The public policies implemented in these years have shown progress in reducing the gap in access to the field of science and technology by women. However, the evidence shows the need to strengthen the monitoring and control processes to ensure their due compliance. Additionally, the government must promote accelerated cultural changes that eliminate gender discrimination in society, being education system a fundamental pillar of change.

Another alternative to strengthen the participation of women in science has been the formation of networks. Support among women has proven to be an effective strategy to overcome discrimination. Women's networks can support their members as well as other professionals who approach the fields of science and technology.

In order to full the objectives of the *Plan Nacional de Desarrollo* ("National Plan for Development"), which aims to generate adequate conditions for the wellbeing of the Ecuadorian people and for the scientific and technological development of country, the inclusion of women within this field should be considered a priority. There is still much work to be done.

6.5 Conclusions

Women in Ecuador and all around the world have traversed a path filled with obstacles in order to be able to advance as autonomous beings capable of making decisions. To this day, this reality has changed and improved; however, it is not the most ideal nor the most equitable. After ten years of implementing different public policies that have improved the conditions to generate science and technology in Ecuador, the stereotypes regarding the role of women have changed very little in much of organizations.

The field of science and technology has been characterized as being highly competitive; however, it is notably less accessible for women. Within this field, there are different shades of discrimination. They manifest themselves with respect to various aspects: salary, harassment, opportunities, knowledge and access.

Sadly, it is fair and necessary to recognize the role that society has played in discriminating against women in this country. In a society as such, where gender inequality is strongly rooted in the culture, it is common to find looks of skepticism when a woman decides that her intellectual capacity and her love for science are the same as those of men. It is difficult and unpleasant to have to face the lack of appreciation for the scientific work of women. It is also difficult to face the fact that they have had to fight against stereotypes on the side in order to win a place within scientific fields and be neither recognized nor accredited for the work they have done.

The discussion about the situation of women within the Ecuadorian university system is an extensive topic that can be discussed from various angles. One of these angles, for example, is the inequality that is demonstrated in the hardships and obstacles that women face, that is, in addition to those that men frequently encounter. The proportion of female researchers and instructors that have been admitted into universities has increased gradually during the last decade thanks to policies and strategies established by the government, which has additionally promoted the increase in female registration in institutions of higher education.

The quality of education in Ecuador has gone through a notable transformation which increased the possibility of including gender perspectives in the university system, understanding that the process of generating and transforming knowledge are fundamental pillars in order to fulfill the principle of quality in education. The polemic regarding the quality of higher education and gender, as a transverse component, has not been analyzed from other points of view, due to which the reduction in the equality gap between men and women should be analyzed critically in order to overcome and eliminate it as soon as possible.

The public policies within the fields of science and technology have changed and, to some extent, improved between the years 2007 and 2017. However, they haven't generated the necessary impact in order to reduce the existing gaps therein. The existing caution between the formulation and implementation of public policies is a problem from which many governments suffer, and it is one of the principal factors for the evaluation of the impact that they have had within the community, in this case, the scientific community and women.

The access to high-level positions for women is an obstacle that requires greater action to face this problem. In Ecuador, the little access of women to positions of power and the subjugation to masculine figures of power at this level are noticeable. There are explicit laws and norms that should be accompanied by changes in conduct and behavior in order to prevent the barriers that women have in their professional lives. There have been significant advances in the related norms, but its implementation still requires a lot of government effort to fulfill the desired goal of equity between men and women.

References

Alianza País. (2006). *Plan de Gobierno del Movimiento PAIS 2007–2011*. Ecuador.

Aquino, M., & Támez, E. (1998). *Teología feminist Latin-American*. Quito: Bayada.

Artigas, M., & Turbine, D. (2007). *El Origen del hombre. Ciencia, Filosofía y Religión*. Pamplona: Eunsa.

Asamblea Constituyente del Ecuador. (2008). *Constitución Política de la República del Ecuador*.

Asamblea Nacional del Ecuador. (2010). *Ley Orgánica de Educación Superior*. Quito.

Ayala Mora, E. (1996). El Laicismo en la historia del Ecuador. *Revista Procesos*, 3–30.

Blanco, M. R. (2005). La educación de calidad para todos empieza en la primera infancia. *Revista Enfoques Educacionales*, 11–33.

Cañedo Andalia, R. (2000). *Ciencia y tecnología en la sociedad. Perspectiva histórico-conceptual*. Habana: ACIMED.

Catton, W. (1980). *Rebasados Las bases ecológicas para un cambio revolucionario*. México: Oceano.

CEAACES. (2014). *Informe General sobre la Evaluación, Acreditación yEvaluación, Acreditación y Universidades y Escuelas Politécnicas*. Quito.

CEAACES. (2016). *Informe de la recategorización voluntaria 2015*. Quito.

Charvet, E. (2016). *Cultura de género y calidad en la universidad ecuatoriana*. Quito: Estudios Estratégicos.

Consejo Nacional de Igualdad de Género. (2014). *Agenda Nacional de las Mujeres y la igualdad de género 2014–2017*. Quito.

Consejo Nacional de Planificación. (2009). *Plan Nacional*. Quito, Ecuador. http://www.planificacion.gob.ec/wp-content/uploads/downloads/2016/02/Decreto-Ejecutivo-1577-2009.pdf

Consejo Nacional de Planificación. (2013). *Plan Nacional del Buen Vivir*. Quito, Ecuador.

Creswell, J. (2007). *Qualitative Inquiry & Research Design (Segunda ed.)*. California: Sage Publications.

Dalla, M., & James, S. (1977). *El poder de la mujer y la subversión de la comunidad*. México: Siglo XXI Editores.

Dufort, L. (2011). *El feminismo de Sor Juana Inés de la Cruz – Lecturas modernas de su Respuesta*. Estocolmo: Stockholm Universitet.

Entrevista 1. (2017, Diciembre). *Mujeres, ciencia y tecnología en el Ecuador* (J. Lin, Interviewer).

Entrevista 2. (2017, Diciembre). *Mujeres, ciencia y tecnología en el Ecuador* (J. Lin, Interviewer).

Entrevista 4. (2017, Diciembre). *Mujeres, ciencia y tecnología en el Ecuador* (J. Lin, Interviewer).

Entrevista 5. (2018, Enero). *Mujeres, ciencia y tecnología en el Ecuador* (J. Lin, Interviewer).

Entrevista 7. (2017, Diciembre). *Mujeres, ciencia y tecnología en el Ecuador* (J. Lin, Interviewer).

Estrada, J. (2016). *Una mujer total, Matilde Hidalgo de Procel*. Lima: Editorial Gustavo Serrano.

García, E. (2008). *Políticas de Igualdad, Equidad y gender Mainstreaming ¿De qué estamos hablando? 33*. España: PNUD.

Goetschel, A. M. (1999). *Mujeres e Imaginarios, Quito en los inicios de la modernidad*. Quito: Abya Yala.

Guardia, S. (Ed.). (2013). *Historia de las mujeres en América Latina*. Murcia: CEMHAL.

Guba, E., & Lincoln, Y. (2000). Paradigmatic, Controversies, Contradictions, and Emerging Confluences. In N. Denzin & Y. Lincoln (Eds.), *The Sage Handbook of Qualitative Research* (pp. 191–216). California: Sage Publications.

Instituto Nacional de Estadísticas y Censos. (2016). *Encuesta Nacional de Ciencia y Tecnología*. Quito.

Jara, R. V. (2014, Diciembre). Apuntes para una reflexión crítica de la política de Educación actual en el Ecuador, desde la visión CTS. Quito: Universidad Politénica Salesiana.

Long, G. (2013). "Suspendida por falta de calidad" El cierre de catorce universidades en Ecuador. In CEAACES (Ed.), *Suspendida por falta de calidad* (pp. 9–30). Quito: El cierre de catorce universidades en Ecuador.

Maiguashca, J. (1994). El proceso de integración nacioanal en el Ecuador: el rol del poder central 1830–1895. In *Historia y región en el Ecuador: 1830–1930*. Quito: Corporación Editora Nacional.

Mora, E. (2013). El paradigma género y mujeres en la historia del tiempo presente. *Revista Historia Autónoma*, 143–160.

Murillo, J., & Martinez-Garrido, C. (2010). *Invetigación Etnográfica*. Madrid: UAM.

Naciones Unidas. (2015). *Declaración Universal de Derechos Humanos*. New York.

Ponce, J. (2016). *Educación superior en Iberoamerica Informe 2016 – Informe Nacional: Ecuador*. Quito: Cinda.

Quecedo Lecanda, R., & Casatño Garrido, C. (2003). Introducción a la metodología de investigación cualitativa. *Revista de Psicodidáctica*, 5–40.

Saldaña, J. (2009). *The Coding Manual for Qualitative Researchers*. Thousand Oaks: Sage Publications.

Sánchez, L. F. (2005). La historia como ciencia. *Revista Latinoaméricana de Estudios Eduactivos*, 54–82.

SCImago. (2018, March 25). *SJR – SCImago Journal & Country Rank*. Retrieved Marzo 27, 2018, from http://www.scimagojr.com

Senescyt. (2015, marzo 6). Retrieved from http://www.educacionsuperior.gob.ec/participacion-femenina-en-la-educacion-superior/

SENPLADES. (2009). *Plan Nacional para el Buen Vivir*. Quito.

Wright de Kleinhans, L., & Alvarado, L. (2005). *Educación y superación femenina en el siglo XIX: dos ensayos de Laureana Wright*. México: UNAM.

CHAPTER 7

Effects of Expenditures in Science, Technology and R&D on Technical Change in Countries in Latin America and the Caribbean

Alexander Cotte Poveda and Clara Carolina Jimenez

7.1 Introduction

This chapter aims to provide an econometric model, using a data panel for Latin America of expenditures in science, technology and research and development (R&D) on technical change. Solow (1957) is a work that has not been examined previously; although many documents discuss the subject, none make the comparison at this level. This chapter will be divided into four parts: the first is the introduction that states the problem, the

A. Cotte Poveda (✉)
Universidad Santo Tomás, Bogotá, Colombia

Colombian Observatory of Science and Technology (OCyT), Bogotá, Colombia

C. C. Jimenez
Colombian Observatory of Science and Technology (OCyT), Bogotá, Colombia
e-mail: cjimenez@ocyt.org.co

C. I. Pardo Martínez et al. (eds.), *Analysis of Science, Technology, and Innovation in Emerging Economies*,
https://doi.org/10.1007/978-3-030-13578-2_7

second is the data and the methodology used, the third is the results and lastly are the conclusions of the model.

Many researchers have addressed the subject of science, technology and innovation, as well as investment in R&D, not only in developed countries but in developing countries as well. In Latin America, one example is Brazil, where Moralles and do Nascimento (2016), in their article on the effects and delays of the indirect effects of R&D in Brazil, refer to government investment in innovation for intensive sectors in science as the capital goods industry. This study was carried out within the framework of a developing economy (Brazil) to estimate the indirect effects on the entire industry of public investment in innovation, as well as the indirect effects of investment in R&D and management made by industrial capital goods throughout the rest of the industrial sector, and the lapse of time between the occurrence of innovative growth of investments and production by account expenses.

The results obtained by these authors' models show significant and positive indirect effects of government R&D expenditures on the capital goods industry with a lapse of three years, as well as a one-year delay due to the occurrence of the effects of the exit by the other companies of the manufacturing sector, the result of innovative investments by the capital goods industry.

On the other hand (leaving the Latin American context for a moment), we have the article by Tuna, Kayacana, and Bektas (2015), whose case study is Turkey and examines the revision of the correlation between expenditure on research and development (R&D) and economic growth at the national level in Turkey, a developed country and the 17th largest economy, which allows the authors to analyze the relationship between economic growth and R&D expenditure. Their analysis found that the inspected series are stationary in the first order and have no cointegrative relationship between them. According to Granger's causality analysis, which they applied, it was also found that there is no causal relationship between the series examined (Tuna et al., 2015).

Zawalinskaa, Tranb, and Płoszajc (2017) discuss the situation of Poland in the European Union and reference the investment goal in R&D, while the document deals with the possible policies focused on this investment to achieve the goal stipulated by the European Union for 2020. It is then proposed (1) to increase government spending on R&D and (2) for taxes to relieve companies in R&D. The method applied to evaluate the options is a dynamic recursive calculation computable general equilibrium model (CGE) for Poland, with an explicit productivity and R&D inventory system.

A slightly closer example in terms of labor production on the subject is Spain, where Acosta, Coronado, and Carlos Romero (2016) conducted research that provides new empirical evidence of the relationship between public support for R&D, spending on the R&D of companies, and innovation and productivity in the food and beverage industry; in this case, a panel from 2008 to 2011 was also analyzed, giving 1910 observations. This research shows that the product of innovation influences productivity with an elasticity of 0.29 for the innovation variable. Other important factors that affect labor productivity include the size of the company, the intensity of the investment, the international competition and the foreign participation of more than 50% in the company (Acosta, Coronado, & Romero, 2016).

Lastly, Anvari and Noriouzi (2016) illustrate the impact of electronic commerce and R&D on the economic development of selected countries and analyze the impact of electronic commerce and R&D, health spending and government size of the gross domestic product (GDP) per capita on 22 countries: Austria, Belgium, the Czech Republic, Denmark, Estonia, Finland, France, Germany, Greece, Hungary, Ireland, Italy, Luxembourg, the Netherlands, Norway, Poland, Portugal, Slovakia, Slovenia, Spain, Sweden and the United Kingdom. The panel model with the Generalized Least Squares (GLS) was used to investigate the 2005–2013 period. The results showed that the explanatory variables in the selected countries played a significant role in per capita income. In other words, it was revealed that e-commerce and R&D expenditure with the GDP per capita had a long-term impact based on the results of the cointegration tests. In addition, it was found that both electronic commerce and R&D have a positive impact on the GDP per capita, but electronic commerce had a greater effect on improving development.

Section 7.2 below discusses the data and methodology. The econometric results are analyzed in Sect. 7.3. Lastly, Sect. 7.4 concludes the study and offers some policy implications.

7.2 Data and Methodology

7.2.1 Data

This section examines the following countries as a representative sample: Argentina, Bolivia, Brazil, Chile, Colombia, Costa Rica, Ecuador, Honduras, Jamaica, Mexico, Nicaragua, Panama, Peru, the Dominican Republic, Salvador and Trinidad and Tobago; the information was captured from World Bank, Ricyt (The Network for Science and Technology

Indicators—Ibero-American and Inter-American), the Colombian Observatory of Science and Technology and the OECD. The period of analysis for the countries in Latin America is from 1999 to 2015.

7.2.2 *Methodology*

We examine the relationships between growth and macroeconomic variables and then between the macroeconomic variables and the changes in both the supplies of factors and the residual, productivity or technical change, in line with Acemoglu (2009), Barro and Sala-i-Martin (2004) and Fischer (1993).

We consider the following production function:

$$Yt = (At,\ Kt,\ Lt,\ Ht,\ It) \tag{7.1}$$

where Y is GDP (see Table 7.1), At is technology, Kt is stock of capital, Lt is labor, Ht is human capital and It is investment. Differentiating (1), we obtain the conventional growth accounting equation.

$$\dot{Y}/Y = \gamma_1 \dot{A}/A + \gamma_2 \dot{K}/K + \gamma_3 \dot{L}/L + \gamma_4 \dot{H}/H + \gamma_5 \dot{I}/I \tag{7.2}$$

where γi is the elasticity with respect to argument i in eq. (7.2). The product $\dot{A}/A$ will be referred to as the technical change.

The motivation for the use of panel data models is to be able to consider in the estimation the effect of unobserved elements (Arellano, 2003; Baltagi, 1995; Hsiao, 2003; Wooldridge, 2002). The unobservable elements are variables that characterize factors of the units or individuals in the analysis, which cannot be measured and therefore cannot be incorporated in the regression model.

Table 7.1 Dependent GDP

	Constant	*Capital*	*Labor*	*Human capital*	*Investment*
Coefficient	−0.741	0.026*	0.950***	0.305***	0.162
Standard error	0.632	0.015	0.029	0.075	0.171
T	1.17	1.75	31.79	4.07	0.94
No. Obs	205				

*, **, *** imply significance at the 10%, 5% and 1% levels, respectively

$$\text{TCit} = \left(\text{S\&T it, R\&Dit, PATit, DR it, SSR it, SCIPUB it}\right) \quad (7.3)$$

where:

TCit = Technical change.

S&T it = Expenditures in Science and Technology made by country i in year t.

R&Dit = Expenditures in Research and Development made by country i in year t.

PATit = Patent application by country i in year t.

DR it = Dependency rate by country i in year t.

SSR it = Self-sufficiency rate by country i in year t.

SCIPUB it = Total science publications by country i in year t.

The structure of the model is as follows:

$$\begin{aligned} \text{TCit} &= \alpha i + \beta\, \text{S\&T it} + \gamma\, \text{R\&Dit} + \upsilon\, \text{PATit} \\ &\quad + \delta\, \text{DR it} + \varphi\, \text{SSR it} + \psi\, \text{SCIPUB it} + \mu it \end{aligned} \quad (7.4)$$

Equation 7.4 represents the basic structure of a model that considers the unobserved elements in which αi are individual or heterogeneous effects and uit are the terms of disturbance. In the model of technical change, the heterogeneous or unobservable effects of the units represent the intrinsic factors of each country that affect the conditions of their productivity.

7.3 Results

We determined the existence of three very common problems in the panel data models: cross-section heteroskedasticity, cross-section correlation and autocorrelation. The above tests show that the estimated residuals of the fixed-effects model have cross-section heteroskedasticity problems, cross-sectional correlation problems and autocorrelation problems (see Table 7.2).

To correct these two problems, a fixed-effects model is estimated with the variance estimator of Driscoll and Kraay (1998). This type of estimation considers the problems of heteroskedasticity and the existence of correlation between groups. In addition, one has the option to correct the existence of autocorrelation of any order. The corrected estimates are shown in Table 7.3.

Table 7.2 Fixed-effects estimations of technical change

Parameter	*[1]*	*[2]*	*[3]*	*[4]*	*[5]*	*[6]*	*[7]*
Constant	−2.171***	−2.341***	−2.299***	−2.422***	−2.395***	−2.363***	−2.245***
	−0.288	−0.383	−0.329	−0.417	−0.324	−0.334	−0.395
Expenditures	0.132	0.119	0.084	0.088	0.032	0.032	0.0004***
in S&T	−0.122	−0.15	−0.144	−0.162	−0.151	−0.152	−0.00018
Expenditures	0.282**	0.259**	0.297**	0.277**	0.409**	0.409**	0.249***
in R&D	−0.111	−0.127	−0.127	−0.14	−0.153	−0.154	−0.075
Number of patents		0.049		0.023			0.08
Applications		−0.079		−0.09			−0.083
Number of			0.023	0.025			0.026
granted patents			−0.034	−0.038			−0.036
Dependency					0.046	0.101	
Rate					−0.039	−0.129	
Self-sufficiency						0.064	
Rate						−0.14	
Science							0.225
Publications							−0.157
σ_υ	0.894	0.898	0.667	0.665	0.933	0.932	0.707
σ_e	0.182	0.187	0.187	0.192	0.183	0.184	0.184
P	0.96	0.958	0.927	0.924	0.962	0.962	0.936
F model	3.03	1.93	6.72	5.12	6.13	7.67	3
Hausman test	0	0	0	0	0.002	0	0.04
Wald test	0	0	0	0	0	0	0
Wooldridge test	0	0	0	0	0	0	0
No. Obs	99	93	87	83	89	89	83

FE Fixed effects. Standard error in parentheses

*, **, *** imply significance at the 10%, 5% and 1% levels, respectively

Table 7.3 Fixed-effects with Driscoll and Kraay standard errors estimations of technical change

Parameter	*[1]*	*[2]*	*[3]*	*[4]*	*[5]*	*[6]*	*[7]*
Constant	−2.171***	−2.341***	−2.299***	−2.422***	−2.395***	−2.363***	−2.245***
	−0.331	−0.252	−0.309	−0.246	−0.353	−0.36	−0.253
Expenditures	0.132*	0.119	0.084	0.088	0.032	0.032	0.0004***
in S&T	−0.073	−0.137	−0.129	−0.177	−0.093	−0.093	−0.00018
Expenditures	0.282***	0.259***	0.297***	0.277***	0.409***	0.409***	0.249**
in R&D	−0.062	−0.069	−0.089	−0.086	−0.105	−0.105	−0.099
Number of patents		0.049		0.023			0.08
Applications		−0.11		−0.11			−0.083
Number of			0.023	0.025			0.026*
granted patents			−0.029	−0.02			−0.014
Dependency					0.046	0.101**	
Rate					−0.034	−0.047	
Self-sufficiency						0.064*	
Rate						−0.044	
Science							0.225***
Publications							−0.055
Prob> F	0	0	0	0	0	0	0
R-squared	0.498	0.483	0.476	0.461	0.493	0.494	0.515
No. Obs	99	93	87	83	89	89	83

FEDK Fixed-effects with Driscoll and Kraay. Standard errors in parentheses

*, **, *** imply significance at the 10%, 5% and 1% levels, respectively

The estimated coefficients of the fixed-effects model corrected for heteroskedasticity and cross-sectional correlation were statistically significant, both individually and jointly at 5% for some of the explanatory variables. In the relation established in the regression analysis, the effect on the technical change is explained by the variables of science and technology, research and development and patents. Therefore, there are country-specific elements that explain the conditions in the technical change and the productivity of each element, which are captured by the individual effects.

The regression results are analyzed using a different approach; in this case spatial interpolation was used. For this interpolation, the algorithm of natural neighbor uses a circular areal-based procedure; this methodology offers more intuitive and simple answers and facilitates the interpretation of the coefficients from both linear and non-linear regressions.

Figures 7.1, 7.2, and 7.3 show the main variables for the selected countries. Figure 7.1a displays two dimensions of the projected surface. Figure 7.1a, which is built with a robust plane, indicates that the plane surface installed in the tridimensional space increases with research and development, and science and technology; thus, the highest part of the surface shows the highest technical change level when the science and technology and research and development levels of countries are increased. Figure 7.1b presents a non-planar relationship between technical change, research and development, and science and technology. In this figure, the highest area represents the highest technical change level associated with the research and development and science and technology levels.

Figures 7.2a-b show that an increase in patents drives a higher growth rate of technical change. Likewise, the countries' science publications level shows the same trend. Therefore, when the patents and science publications levels are higher, the technical change growth rate should be higher.

Figures 7.3a-b show the results of this analysis, suggesting that factors such as education, expenditure in technology and communications influence the total factor productivity and economic growth and positively impact economic and social development in countries.

7.4 Conclusions

In this chapter, we attempted to explain and analyze technical change and the relationship with objective conditions using a data panel with fixed-effects in countries in Latin America and the Caribbean between

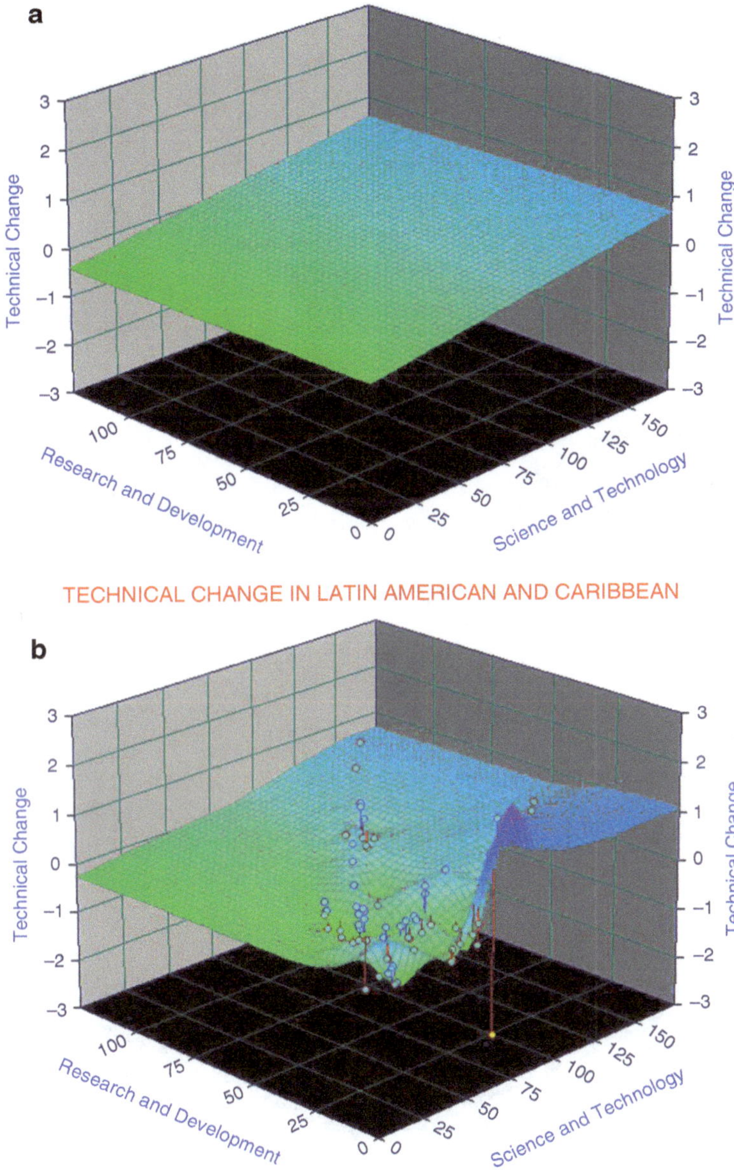

Fig. 7.1 Technical change, research and development, and science and technology in Latin America and the Caribbean

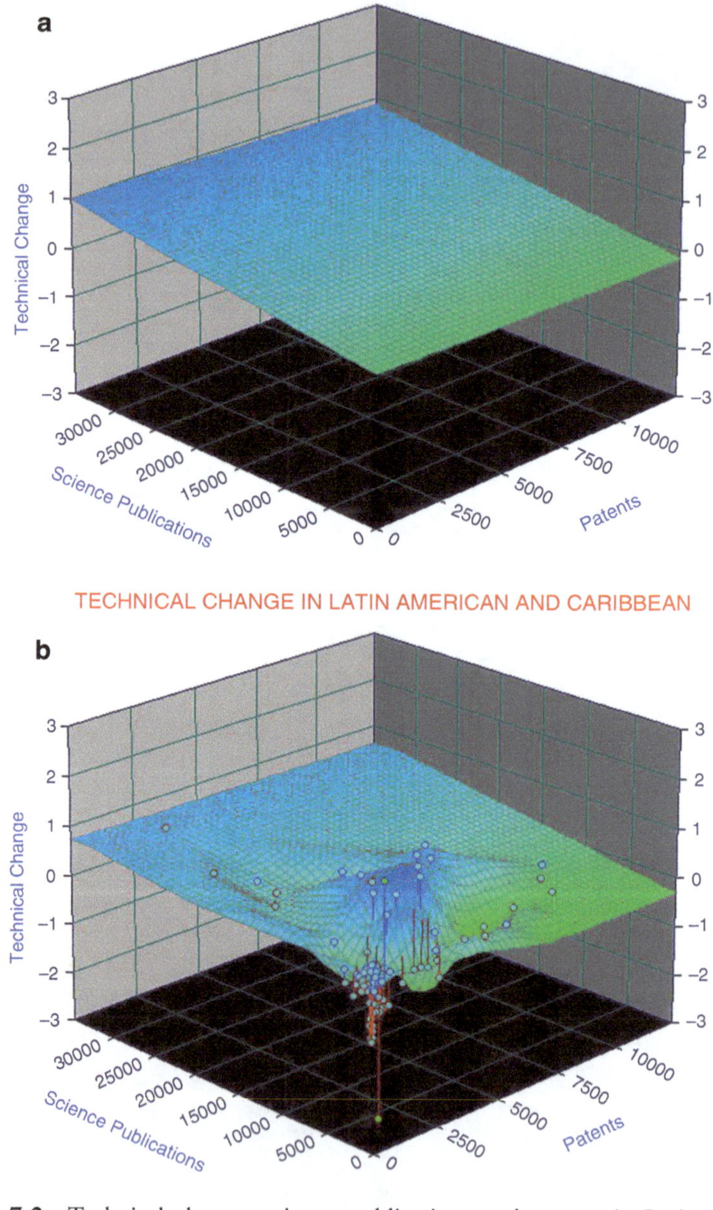

Fig. 7.2 Technical change, science publications and patents in Latin America and the Caribbean

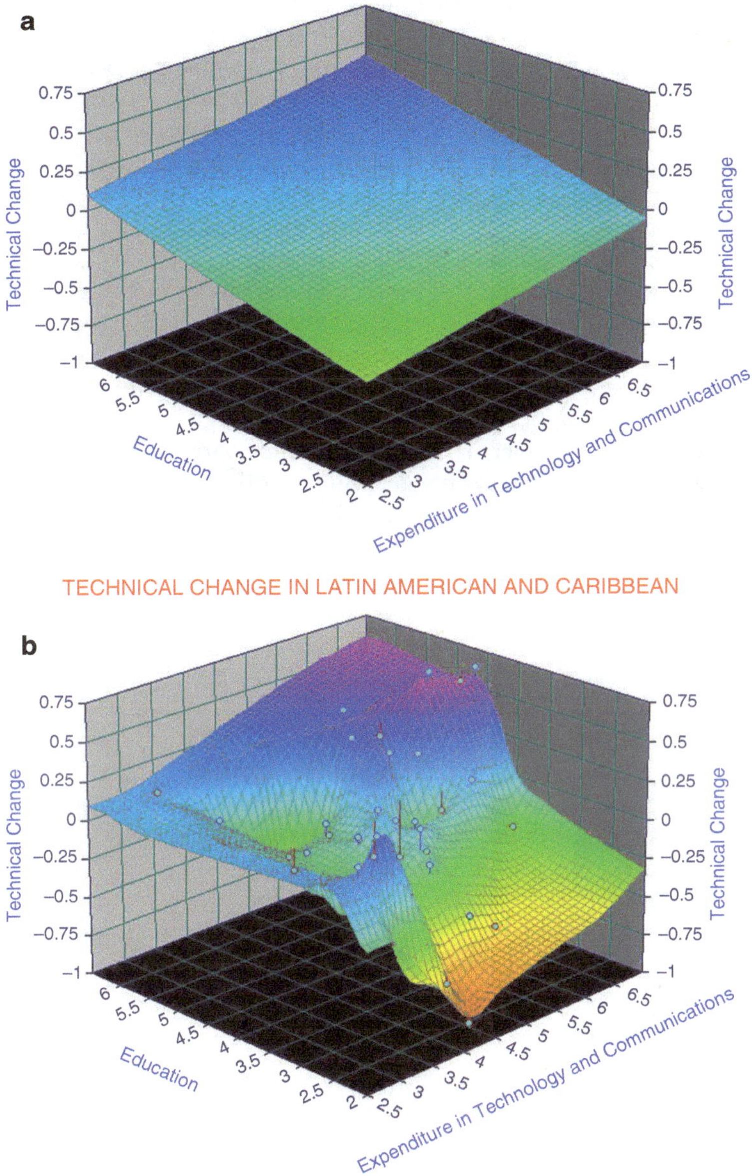

Fig. 7.3 Technical change, education and expenditure in technology and communications in Latin America and the Caribbean

1999 and 2015. Several studies have analyzed the relationship among technical change, expenditures in science, technology and R&D and socio-economic development from a two-dimensional perspective. In this analysis, technical change, expenditures in science and technology, expenditures in research and development, patent application, dependency rate, self-sufficiency rate, science publications by country and socio-economic development form the framework using a panel with fixed-effects techniques for several countries, with the purpose of identifying connections among these variables.

The econometric results suggest that an increase in the expenditures in S&T, R&D and education increase technical change measured as total-factor productivity (TFP). On the other hand, an increase in patents, measured by the number of patent applications and the number of granted patents, leads to a higher growth in technical change in countries in Latin America and the Caribbean. Surprisingly, some variables have no significant relation with technical change, whereas other variables were significant. When performing corrective actions with fixed-effects using Driscoll and Kraay standard error estimations, most of the variables of interest were significant.

From a policy perspective, science, technology and innovation policies should promote key strategies to guarantee economic growth and development and to decrease poverty through higher equality, better employment conditions and quality education that generate higher income and provide greater well-being for the population.

The findings demonstrate the importance of generating social policies and strategies to decrease the gaps in science, technology and innovation and to increase economic growth and development, productivity and investment for those countries. These policies and strategies should include investments in education while increasing opportunities in the labor market, strengthening the justice system and generating an effective allocation of resources in Science, Technology and Innovation (STI).

References

Acemoglu, F. (2009). *Introduction to Modern Economic Growth*. Cambridge, MA: MIT Press.

Acosta, M., Coronado, D., & Romero, C. (2016). Linking Public Support, R&D, Innovation and Productivity: New Evidence from the Spanish Food Industry. Food Policy, 309–325.

Anvari, R. D., & Noriouzi, D. (2016). The Impact of E-commerce and R&D on Economic Development in Some Selected Countries. *Procedia, 229*, 354–362.
Arellano, M. (2003). *Panel Data Econometrics*. Oxford: Oxford University Press.
Baltagi, B. H. (1995). *Econometric Analysis of Panel Data*. Chichester: Wiley.
Barro, R., & Sala-i-Martin, X. (2004). *Economic Growth* (2nd ed.). Cambridge, MA: MIT Press.
Driscoll, J. C., & Kraay, A. C. (1998). Consistent Covariance Matrix Estimation with Spatially Dependent Panel Data. *Review of Economics and Statistics, 80*, 549–560.
Fischer, S. (1993). The Role of Macroeconomic Factors in Growth. *Journal of Monetary Economics, 32*(3), 485–512.
Hsiao, C. (2003). *Analysis of Panel Data*. Cambridge: Cambridge University Press.
Moralles, R. H., & Nascimento, D. A. (2016). The Effects and Time Lags of R&D Spillovers in Brazil. *Technology in Society, 47*, 148–155.
Solow, R. (1957). Technical Change and the Aggregate Production Function. *Review of Economics and Statistics, 39*, 312–320.
Tuna, K., Kayacana, E., & Bektas, H. (2015). The Relationship Between Research & Development Expenditures and Economic Growth: The Case of Turkey. *Procedia, 195*, 501–507.
Wooldridge, J. (2002). *Econometric Analysis of Cross Section and Panel Data*. Cambridge, MA: MIT Press.
Zawalinskaa, K., Tranb, N., & Płoszajc, A. (2017). Science Direct R&D in a Post Centrally-planned Economy: The macroeconomic Effects in Poland. *Journal of Policy Modeling*, 1–23.

CHAPTER 8

Science, Technology and Water Access

Gustavo Correa Assmus

8.1 Introduction

The history of Colombia's economic and social growth is closely linked to its natural richness, which is currently highly affected by processes of degradation, extinction, pollution and fragmentation of ecosystems. There is no doubt that Colombia's national water supply is one of the most important worldwide, which may explain why its management and protection are not yet assured with the sociopolitical and cultural responsibility that merits a natural resource on which life depends.

Despite the legislative advances, programs and investments that are carried out, the environmental degradation that consumes approximately 4% of the annual gross domestic product (GDP) (Correa, 2016; Sanchez, Ahmed, & Awe, 2007) does not feature deceleration and efficient management, but it does generate significant impacts on the poorest and most vulnerable population.

The lack of coordinated water policies with sustenance in science and technology (S + T) does not provide a framework of sustainability for the important task of conserving and properly managing the natural supply of

G. Correa Assmus (✉)
Universidad de La Salle, Bogotá, Colombia

C. I. Pardo Martínez et al. (eds.), *Analysis of Science, Technology, and Innovation in Emerging Economies*,
https://doi.org/10.1007/978-3-030-13578-2_8

water, as a fundamental input for the development of the population, the environment and the nation.

Events such as the depletion of water resources, their increasing and systematic contamination or the lack of allocation mechanisms denote the state's lack of interest in matters of national water management and governance, a phenomenon that encourages economic interests to pursue the privatization of water and consolidates power figures that do not recognize socio-environmental rights and duties.

If we add to this phenomenon the challenges of climate change and the occurrence of other impact phenomena such as the El Niño–Southern Oscillation (ENSO), we understand the need to pursue an advance in science and technology for access to water (S + T + H_2O), with the purpose of granting some degree of sustainability to the innovation and development (R + D) of the water base required by the country in order to establish paths respectful of natural resources and that, in turn, contribute to the advancement in the quality of life and the generation of national wealth. Hence, the objective of this chapter is to present an approach to some shortcomings in science and technology (S + T) that presents the water subsector.

However, the results and conclusions show a weak condition in the field of building knowledge that allows to us consolidate a subsector of water, robust against the conditions, to articulate social, economic and ecological relations in context. The presentation of the theme begins in the theoretical framework, followed by the method of study used, which leads to the results and discussion that allow us to establish some conclusions.

8.2 Theoretical Framework

While it is true that efforts to improve coverage and access to water in Colombia have increased, especially since the launch by the United Nations of the Millennium Development Goals (MDGs), consolidated since 2015 as the Sustainable Development Goals (SDG), there is still work to be done in this field. The trend in the coverage of the aqueduct at the national level has a broad bias in favor of the urban supply over the rural supply. The official information reports a coverage slightly higher than 90% of citizens nationwide, logically measured from the urban population concentration. However, rural access is less than 40%, a situation that leads to a national average coverage of approximately 65% (Correa, 2016; DANE, 2013).

It is clear that the dispersed population in Colombian rural areas is not a priority for access to water, at least in quality and quantity, mainly due to costs, technical demands and interest in a short-term investment recovery, besides other factors. Therefore, their right to water is violated due to the low importance given to the issue by both public and private organizations. If the government interest in yielding the responsibility for hydric social coverage to private entities is added to these issues, the problem can fall into the sphere of the little recognized[1] cases, also known as "invisible".

The interest to modernize the sector is oriented preferentially to the bureaucratic consolidation of organizations, not to the scientific and technological development that allows us to make an offer in quantity and quality of the resource under considerations of equity and ecological sustainability. For this reason, the Ombudsman's Office (2009, 2012) draws attention to 3% of the municipalities (large cities) with drinking water suitable for consumption, 6% with medium risk (the main intermediate cities), 73% with water not suitable for human consumption (headwaters and villages) and 18% with unviable water (ruralities and scattered ruralities). These data indicate structural deficiencies in the background, with regard to a service directly linked to a human right.

The issue can be further exacerbated if we take into account that climate change is taking place with a gradual but steady increase in global temperature, with which a worsening of the phenomenon of water scarcity is foreseeable in the future, from which Colombia it is not exempt[2] (PNUD, 2008). Moreover, generally, the resources are available once the shortage has occurred; thus, there is no preventive policy with social responsibility. Hence, it should be taken into account that wastewater, once it enters the distribution networks, varies between 25% and 75% depending on the city, mainly due to the age and design deficiencies (Defensoría del Pueblo, 2010).

History has shown that the attempt to implement depoliticized corporate models for the administration of natural resources, especially water, has been a failure that nurtures social perceptions about the inability of the state to establish integration mechanisms and institutional control. In this order of ideas, the state promotes the privatization of the water coverage

[1] The situation supported by the document of the National Council of Economic and Social Policy (CONPES) 3383 of 2005.

[2] It is worth remembering the situations of drought presented at different times, in Departments such as Guajira, Boyacá, Meta, Casanare or San Andrés, among others.

service but with resources from the national budget, which generates a phenomenon of corruption reflected in inoperable aqueducts, investments that were never carried out and another series of anomalies affecting the need of the inhabitants to have access to water (DNP, 2005).

The process of modernization (essentially bureaucratic) continues with the privatization phase of the service, where the lending entities are responsible for their investments. This situation has increased citizen vulnerability not only because of the climate impact but also because of the business impact of low private investment to respond to citizen demands. This situation is reflected in continuous or long-term cuts, low water quality offered for consumption, or simply not providing the service, among other events. In addition, inequality in land tenure (GINI = 0.8) has favored a growing inequality in the possession of water, via the concentration of resources in owners and for-profit organizations that do not respect the social right to water (Correa & Muñoz, 2015).

From the perspective of science and technology (S + T), in terms of building useful knowledge for universal access to water in quantity, quality and economy, the processes of creation and innovation are not the most visible. In a wide range, investments are oriented toward the import of technology (construction machinery) with which related companies can structure an offer that allows them to maintain or increase profit margins as independent organizations that face a market.

Therefore, science and technology for access to water (S + T + H_2O), accounts for the delay over time that scientific research maintains in different areas of national development, but in this particular case, has a greater gravity due to several aspects. First, access to water is an inalienable right of citizens for whom the state is directly responsible. Second, historically, the natural water supply has been generous in Colombia, but currently, there are recessive processes that cause constraints in different regions of the country. Third, climate change has an impact in Colombia due to the equatorial location of the country; this is an important condition for mitigation.

8.3 Data and Methods

The sources of information taken into account at the governmental level are the Administrative Department of Science, Technology and Innovation (COLCIENCIAS), National Planning Department (DNP), Colombian Observatory of Science and Technology (OCyT), National Administrative

Department of Statistics (DANE), Potable Water and Basic Sanitation Regulatory Commission (CRA), Ombudsman's Office, Institute of Hydrology, Meteorology and Environmental Studies (IDEAM) and Ministry of Health and Social Protection (MINSALUD). From the private sector, studies related to the subject were taken into account, carried out at La Salle University, Javeriana University, the United Nations Development Program (UNDP), the World Bank (WB) and the Organization for Cooperation and Economic Development (OECD).

The empirical databases were constructed from absolute values for the period of 2005–2015, in order to subject them to different exercises that allow the construction of interpretations regarding the proposed topic. Consequently, the analysis of marginal variation is characterized by establishing the real behavior between the consecutive values assumed by the variable under study (Samuelson & Nordhaus, 2002), where its application is established as follows:

$$VM = \left[(Xi+1) - (Xi) / (Xi)\right] * 100 \tag{8.1}$$

Given that the databases have been constructed with nominal values and it is necessary to evaluate their status using real formulas, it is necessary to deflate their initial values based on the consumer price index (CPI) in order to eliminate the inflationary effect on the items (DANE, 2010); thus, we obtain equation (8.2):

$$D = (Xt_1) - (IPCt_1) \ldots\ldots Ntn \tag{8.2}$$

The dependence of the items is based on the origin of the resources, so the correlation with variables such as gross domestic product (GDP) or taxation (T) merits the application of the correlation coefficient, as follows in equation (8.3):

$$r = \sqrt{}SCR / SCT \tag{8.3}$$

where SCR measures the deviation around the mean in the reference variable (Y) for the portion of it explained in the regression model (SCT). As explained above, this is done in order to establish the strength of the relationship between the intervening variables (X, Y) (Webster, 2000).

The validation of the results obtained from the empirical data and the predictions shows coherence with the characteristics, behaviors and attributes observable in the economic and daily reality. This validation confirmed that the results are not a mere coincidence or phenomenon that randomly occurred from 2005 to 2015 (Thornley & France, 2007).

8.4 Results

To take into account the different analyses obtained from the study, the results ordered by subject are presented below and some of them include figures that allow for knowing the behavior of the variables under study for the years taken into account.

8.4.1 Business Characterization

Colombia has a total of 32 departments and 1102 municipalities; overall, a total of 140 public or private companies participate in the process of capturing and distributing water.[3] These companies operate in 215 municipalities of urban preference that are located in the central part of the country, given its greater population, which makes the provision of service more profitable. Regarding these companies, the majority of them are constituted as administrative boards, aqueduct and sewage companies, user associations or community action boards. Additionally, the water they use to fulfill their service usually comes from rivers, streams, springs, springs, wells, cisterns or aquifers (Ministry of Health and Social Protection, MINSALUD, 2016).

From another perspective, it is estimated that for companies in the catchment, water purification[4] and distribution sector are innovative in 29.2% of them, the innovation potential is 4.5%, and 66.2% are not innovative. Their interests in innovation are mainly oriented to strengthen their commercial management in 10.9%, consolidate their internal organization in 19.4% and improve the supply service in 69.6%. This service improvement (69.6%) is sought through acquiring imported machinery and equipment by 54.2%, acquiring information technology and using communications in the office by 27.4%, offering technical assistance to the

[3] Organizations that distribute water in tank trucks, or aguateros, that supply some marginal poor communities are not included.

[4] It should be noted that not all these organizations assume the purification of water for consumption.

user by 18.4%, and the number for the research and development (R + D) budget allocation is not available (National Administrative Department of Statistics, DANE, 2016).

8.4.2 *Budget Participation*

This study compares R + D items in relation to GDP and S + T in relation to the budget of the water subsector, the behavior of the trends is quite similar; therefore, its correlation is high and positive ($r = 0.7573$), with an index of significance at 1%, which indicates that the relationship between the budget of science and technology for development in particular of the water subsector is quite dependent on the behavior of historical GDP.

The most relevant aspect is the low participation rate in the distribution of resources, which can translate into little institutional and eventually social interest, due to scientific progress in general and in particular of the water subsector, in terms of management, distribution and conservation, for the benefit of national development.

8.4.3 *Inflationary Effect*

The behavior of the budget for the subsector of water in both nominal and deflated terms shows a very similar behavior, which indicates that its annual adjustment rate has a close relationship with the inflationary behavior; thus, over 11 years, the average budget reduction due to inflationary impact was approximately 4.22%.

8.4.4 *Tax Participation*

The participation of the specific budget for (S + T) in the total amount of taxation collected by the DIAN (2017) is relatively low; its variation rate oscillates through the 11 years between 2.27% and 5.08%, with an average of 3.39%. However, participation has been decreasing since 2010 until reaching 2015 levels that were not observed seven (7) years ago, which means that their current level of participation is similar to that of 2008.

8.4.5 *Training of Scientific Staff*

For 2015, of the total programs at the level of master's degree offered in the country for higher education, only 2% correspond to the field of engineering and technology, a category in which the training is related to the management of water. Thus, the situation with regard to doctorates denotes a greater insufficiency because their participation is reduced to 1.7%. From the perspective of the students, the interest in studying these subjects does not show an improvement in their indicators; for the same year of 2015 of the total of students in masters, 1.14% were in engineering and technology, while 1, 07% are PhD students (OCyT, 2016). Therefore, the supply of qualified labor to intervene in the environment is low and, for its part, the sector is not projected to be capable of the scientific and technological renewal of it, which is why job placement spaces are relatively scarce.

One aspect to consider is the process of contraction that has been presented in the number of researchers active in all knowledge. For 2006, there were 14,710 researchers, which decreased to 6036 in 2015, indicating an average annual reduction rate of 5.90%. Similar results were observed in the number of researchers linked to groups: 12,912 researchers in 2006, 5813 researchers in 2015 and an annual average decrease rate of 5.50% (OCyT, 2016). These findings suggest that there is little interest in the knowledge of the qualified personnel by which all the knowledge and the perspectives of improvement in (S + T + R) are affected.

8.4.6 *Patent Development and Innovation*

The gap between patents requested by residents and non-residents, nationally via the Superintendence of Industry and Commerce (SIC) and those finally granted is significant, which indicates a possible waste of efforts and resources by not taking into account the rigor demanded by development is the field of (R + D). Although it is true that the register has been growing since 2009, the evolution of patents granted varies with a downward trend toward the final years of the period under study. It should be noted that, to a large extent, the contribution of patents for the water subsector is limited by the lack of qualified personnel, analyzed in the previous topic.

8.5 Discussion

It is clear that before access to water, the most vulnerable see their rights restricted to interests that exceed their voices (PNUD, 2006). The situations of inequity with the water resources generally reproduce poverty, harm citizen health and increase social, economic and ecological costs, all together with processes that undermine human dignity.

Therefore, the (S + T + H_2O) must assume responsibility and promptly manage the resource in order to guarantee quality for consumption, contribute to its equitable distribution, reduce human and ecosystem vulnerability in a context of climate change and propose opportunities for innovation and development (R + D) coherent with territorial capacities.

From a territorial perspective, it should be noted that the negative impact on water conservation, that is, waste or its contamination on lentic, lotic or aquifer sources, has implications on the natural water supply that has in many cases gone unnoticed, whose effects are generally delayed (IDEAM, 2010). Therefore, the anthropogenic water footprint must take into account in its control the integral advances in (S + T) that involve interdisciplinary conditions and multipurpose solutions, which are participative and respectful of the environment (IDEAM, 2014).

The similarity in the behavior of the budget evolution analyzed by means of an indicator for (S + T) and another for (R + D) supposes that the budgetary allocation does not strictly obey a concept of development in science, technology and innovation (S + T + R) but rather a sectoral allocation in accordance with history. Thus, it would not be responding to development needs but to pre-established budgetary requirements.

These findings suggest that there are insufficient manifestations of governability (Hemson, Kulindwa, Lein, & Mascarenhas, 2008) because in the water sector the participation of (S + T) in the monetary mass and budgetary resources are limited.

The low projection in (S + T + H_2O) generates sub-optimal organizational structures that result in poor or regular quality services, as stated in the theoretical framework, which, in turn, finds an echo in poor governance that leads to anomalous conditions in the disposition of the resources and the expected results from society (SUPERSERVICIOS, 2010).

On one hand, it can be argued that given the economic conditions of the country and the requirements that poverty demands, it does not seem that the (S + T + H_2O) is a high-impact item for short- and medium-term government policies. It is also questionable to try to achieve citizen welfare and build opportunities behind the access to water, not only as a socioeconomic and ecological option but also as an inalienable right of citizens. At this point, it makes sense to discuss the limited expectations in terms of Krugman (2016) and propose changes with which the desired country and society can be defined, in the face of an increasingly globalized, competitive future affected by climate change.

8.6 Conclusions

The evidence herein indicates that many of the municipalities of the country lack a regulating and distributive entity or organization of the water service, thus revealing the absence of the state and the low private interest in consolidating the development of the subsector. This situation can be explained from the economic view of the situation. Costs and the profitability of the service keep a significant number of inhabitants away from urban centers or dispersed in rural areas, thereby marginalized from access to water as a right.

Most of the companies or organizations providing the service are not innovative (66.2%), and their budget destined for (S + T) is not statistically significant; therefore, the (R + D) indicators are deficient for the subsector. From the official budget allocation for (S + T), two considerations can be established: the first is that budgets maintain a high dependence on the historical behavior of both GDP and inflation. Second, their time series denote a low real participation in the monetary allocation. Therefore, a good part of the expectations of the (R + D + H_2O) in the field of management, water conservation and the mitigation of climate change falls on institutions such as universities or research centers with an interest in the subject but also with limited resources.

For its part, the formation of human capital with scientific and technological capabilities related to water treatment, as well as the innovative contribution to development seen from the generation of patents, is not due to a sustainable policy that can benefit both the water sector and its users. This is an important situation, given the importance of the water resource to sustain the progress of the nation and the need to mitigate or adapt to climate change.

References

Correa, G., & Muñoz, A. (2015). *Agua, pobreza y equidad, un análisis asimétrico.* Bogotá: Bioética. Universidad Militar Nueva Granada.

Correa, G. (2016). *Agua, pobreza y desarrollo en Colombia, un análisis para el periodo 1970–2012.* Bogotá: Universidad de La Salle.

DANE. (2010). *Metodología para el cálculo de índices deflactores.* Bogotá: Dane.

DANE. (2013). *Encuesta nacional de hogares.* Bogotá: Dane.

DANE. (2016). *Encuesta de desarrollo e innovación tecnológica del sector servicios.* Bogotá: Dane.

DEFENSORIA DEL PUEBLO. (2010). *Diagnóstico de la accesibilidad económica y el acceso a la información.* Bogotá: DESC.

DIAN. (2017). Estadísticas de los ingresos tributarios. Bogotá: Coordinación de Estudios Económicos.

DNP. (2005). *Seguimiento a la participación del sector privado en acueducto y alcantarillado realizada desde el programa de modernización empresarial.* Bogotá: DNP.

Hemson, D., Kulindwa, K., Lein, H., & Mascarenhas, A. (2008). *Poverty and Water. International Studies in Poverty Research.* New York: CROP.

IDEAM. (2010). *Estudio Nacional de Aguas ENA.* Bogotá: IDEAM.

IDEAM. (2014). *Estudio Nacional de Aguas ENA.* Bogotá: IDEAM.

Krugman, P. (2016). *La era de las expectativas limitadas.* Bogotá: Ariel.

MINSALUD. (2016). *Informe nacional de calidad del agua para consumo humano (INCA – 2015).* Bogotá: MINSALUD.

OCyT. (2016). *Indicadores de ciencia y tecnología 2016 Colombia.* Ed. Ántropos. Bogotá.

Ombudsman's Office. (2009). *Annual Report.* Toronto: Office of the Ombudsman.

Ombudsman's Office. (2012). *Annual Report.* Toronto: Office of the Ombudsman.

PNUD. (2006). Más allá de la escasez: poder, pobreza y crisis mundial del agua. Informe sobre el desarrollo humano 2006. Mundiprensa. Madrid.

PNUD. (2008). La lucha contra el cambio climático: solidaridad frente a un mundo dividido. Informe sobre el desarrollo humano 2007–2008. PNUD. Madrid.

Samuelson, P., & Nordhaus, W. (2002). *Economía* (17th ed.). Madrid: McGraw Hill.

Sanchez, E., Ahmed, K., & Awe, Y. (2007). *Prioridades ambientales para la reducción de la pobreza en Colombia.* Bogotá: Banco Mundial.

SUPERSERVICIOS. (2010). Estudios Sectoriales de Acueducto, Alcantarillado y Aseo 2006–2009, Superintendencia de Servicios Públicos Domiciliarios. Bogotá.

Thornley, J., & France, J. (2007). *Mathematical Models in Agricultural, Quantitative Methods for the Plant, Animal and Ecological Sciences.* London: CABI.

Webster, A. (2000). *Estadística aplicada a los negocios y la economía* (3rd ed.). Bogotá: McGraw Hill.

CHAPTER 9

The Ecuadorian Scientific Technological Policy Within the Framework of "Buen Vivir", a Dispute with the Orthodoxy of Development

Fernando Herrera García, Jessica Lin Gutiérrez, and Antonio A. Franco-Crespo

9.1 Introduction

In Latin America, the first decade of the new millennium is characterized by a "turn to the left" (post-neoliberalism) by most governments (Stoessel, 2014). Above local particularities, post-neoliberalism is the intention of overcoming the postulates of the Washington Consensus and by the recovery of state abilities in the handling of the economy and in public agenda. The post-neoliberal alternative in Ecuador, the citizens' revolution, came to power in 2007 under President Rafael Correa and posed the goal of achieving a *Buen Vivir* or *sumak kawsay*, a "radical break" (Ramírez, 2013)

F. Herrera García (✉) • J. Lin Gutiérrez • A. A. Franco-Crespo
Department of Organizational Studies and Human Development, Escuela Politécnica Nacional, Quito, Ecuador
e-mail: edwin.herrera@epn.edu.ec; jessica.lin@epn.edu.ec; antonio.franco@epn.edu.ec

C. I. Pardo Martínez et al. (eds.), *Analysis of Science, Technology, and Innovation in Emerging Economies*,
https://doi.org/10.1007/978-3-030-13578-2_9

or "re-foundational" with development orthodoxy. Streeck and Thelen (2005) call it "conversion". That is, existing institutions are adapted or "redirected to new goals, functions and purposes" (Streeck and Thelen, 2005, p. 26), depending on the interests of new actors. In Ecuador and other post-neoliberal Latin American countries, "the debate on the relationship between three dissociated participants during the peak of neoliberal policies: science, the state and the production has been revitalized" (Postolski & Funes, 2015, p. 11).

We propose the study of the change in Ecuadorian scientific technological policy (PS&T) based on the role of scientific community and the political elites, or "political cultures" (Elzinga & Jamison, 1996). The analysis focuses on changing the conceptual perspective of scientific technological policy (PS&T), omitting the study of the definition and implementation of PS&T or its instruments. In addition, in Ecuador research activities have been concentrated mainly in a few universities and to a lesser extent in a few public research institutes. Before speaking about the scientific community, it is more appropriate to speak of an academic community made up of teachers with third-level training from the largest, oldest and most prestigious universities in the country, who spent part of their time to research.

These microsocial processes are analyzed in the social, political and economic contexts of Ecuador. In addition, since PS&T is also due to internationalization that can be explained through the role of international organizations and multilateral bodies, Vélho (2011) scientific and technological paradigm is used. It is interesting to investigate which conceptual elements characterize the change of PS&T of Ecuador beginning in 2007, what conditions or concrete processes determined the change in PS&T, who is involved and what their ideas are, how they have intervened, what relations or tensions have existed between them and to what extent PS&T has assumed international models or paradigms.

9.2 Literature Review

The best translation *sumak kawsay* is "life in fullness" (Huanacuni, 2010) and in the Ecuadorian constitution implies a heterodox conception of development whose objectives are to improve the people's quality of life and develop their capabilities and potential, have an economic system that promotes equality through social and territorial redistribution, promote the effective participation of citizens in all areas of public interest, establish harmonious coexistence with nature, ensure national sovereignty, promote Latin American integration, and protect and promote cultural diversity (Ecuadorian Constitution 2008, article 276). Under this theoretical paradigm, the Correa's administration proposed to restructure society's belief system and modify the "Ecuadorian institutional system" (Ramírez, 2010, p. 133) as well as the production of public policies (Palacios, 2014, p. 16). In this sense, "*Buen Vivir*" can be understood as a new global referential of Ecuadorian public policies, "the set of prescriptive regulations that give meaning to a political program, by defining selection criteria and modes of designation of objectives" (Muller, 2009, p. 117). As in the other sectors of state intervention, this determines the definition of PS&T.

Several authors (Dickson, 1984; Elzinga & Jamison, 1996; Powell, Owen-Smith, & Smith-Doerr, 2011; Sanz, 1997, 2001; Whitley, 2010) have studied the evolution and change of PS&T, beginning with the participants' interaction. Among these theoretical referents, we opted for the concept of "political cultures" of Elzinga and Jamison (1996), in force for over 20 years and useful in a developed country or in countries with emerging economies, such as Ecuador. Its basic contribution is identifying the relevant participants, their ideas and interests and expected behaviors in the development of PS&T.

Its approach is that changes in PS&T are explained by the interaction of the management elites, which are bearers of what they call "political cultures". They consider that the balance between cultures and the evolution of their relative weights marks the stages of PS&T evolution. Political cultures, such as bureaucratic, academic, economic and civic, bring together the most relevant participants in science and technology (S&T), each with its own interests, ideologies, institutional bases, traditions and relations with political and economic power. The bureaucratic culture is based on the administration of the state and believes that science is important for its social use, "science for politics". On the other hand, academic culture focuses on "politics for science" and on preserving the values of the scientific community: autonomy, integrity, objectivity and control of investment and organization. Economic culture is related to business and prioritizes the technological uses of science for its dissemination in the market. Civic culture is related to social movements and is concerned with the consequences and implications of S&T. However, in our research, these participants will be reduced to two: academic and bureaucratic, due to the lack of interest that entrepreneurs historically have had in the development of S&T and to the limited impact that civil society has had on the development of Ecuadorian PS&T.

On the other hand, several authors (Averch, 1985; Blume, 1985; Bonvillian, 2011; Brooks, 1982; Dickson, 1984; Elzinga, 1988; Elzinga & Jamison, 1996; Freeman, 1987; Marí, 1982; Rip & Hagendijk, 1988; Rothwell & Zegveld, 1985; Ruivo, 1994; Salomon, 1977; Vélho, 2011) have analyzed the evolution of PS&T. Its categorization has depended on the approach used (political, institutional, economic criteria, among others), the various aspects of the policy studied or geographical contexts. One of the most suggestive and comprehensive proposals is Vélho's (2011), who studies the development of PS&T based on a scientific technological paradigm. She considers there is great similarity in periodization and descriptions of the evolution of explicit PS&T in different countries, and that this similarity has persisted over time despite the changes in PS&T. This similarity is due to internationalization of PS&T through the dissemination of normative models by various international organizations after World War II.

Vélho (2011) defines each paradigm according to the dominant science concept; the participants that produce knowledge; the relationship between science, technology and innovation (STI); the rationality and focus of politics; and instruments of analysis and evaluation. She considers four paradigms: (1) "science as an engine of progress" (postwar—early 1960s) conceives science as neutral, universal and self-regulated, where knowledge is produced by scientists (science republic); STI relationship is linear and "technology push"; the rationality of PS&T is in strengthening research and promoting offer; the focus is on scientific policy; and input and peer evaluation indicators are used; (2) "science as a solution and cause of problems" (1960s–1970s) considers science to be neutral but must be controlled; knowledge is produced by scientists in contact with demand; that the STI relationship is linear and "demand pull"; rationality lies in the identification of priorities and in the offer-demand relationship; focus is on technology policy; and uses outcome indicators and peer review; (3) "science as a source of strategic opportunity" (1980s, 1990s and early 2000s) says science is socially constructed; considers knowledge is produced by scientists and engineers influenced by a complex network of participants; the STI relationship is based on interactive models that integrate offer and demand; the rationality of politics is based on strategic programs, collaborative research and co-participation; focus is on a policy of innovation; and uses extended peer review, impact analysis and forecasting; and (4) "science for the good of society" (twenty-first century) which is still speculative and based on a moderate constructivism questioning the imitation of policies, prioritizing national styles and assuming a path dependent and local knowledge, and which assumes that knowledge is produced by a network of participants in diverse configurations depending on the circumstances. This considers that the STI relationship is based on interactive models and social choice, which assumes that rationality lies in coordination and management and on an independent scientific basis; focus is on welfare policy, and evaluation is based on citizen participation, the building of scenarios and ex-ante evaluation.

An approach in the new institutionalism or political economy is used; it considers that "public policies are not simply objects of choice for a social planner who tries to maximize the welfare of the population", but rather "they emerge from a decision-making process that involves a multiplicity of political participants interacting in a variety of scenarios" (Scartascini, Spiller, Stein, & Tommasi, 2011, p. 2). As suggested by Sanz (1997), this approach and the analysis of public policies explain the development of PS&T, focusing attention on the role of ideas, interests and institutions. Additionally, it allows us to consider macrosocial processes in PS&T development, and microsocial processes of formulation, design and implementation of these policies.

An analytical narrative accounts for the development of Ecuadorian PS&T. Becker (2009) explains it is a perspective of neo-institutional analysis focusing on narrating "processes" that explain what we want to understand, and how it has become what it is. It is assumed that what happened does not have a cause, but rather a story or narrative. To understand events of interest, we investigate the process that leads them to happen instead of the conditions that made them necessary. The facts are caused only by the history that led them to be as they are. The analytical narrative aims to "locate and explore specific mechanisms that shape the interaction between strategic participants and thus generate results" (Geddes, 2006, p. 38). It implies "first, extracting from the narrative the key actors, their objectives, and their preferences and the effective rules that influence their behavior;" and "second, to create strategic interactions that produce a balance limiting some actions and facilitating others" (Geddes, 2006, p. 39). The main aspect is to identify why there is change from one institutional equilibrium to another over time (Levi, 2006, p. 14).

9.3 Methodology

The research proposed is qualitative, a case study centered on the PS&T change begun in 2007. Research techniques used were documentary analysis and in-depth interviews.

The legal framework of PS&T, the regulations implemented by S&T formulation, planning and promotion bodies were analyzed. The discourses contained in the Constitution of 2008, the Development Plan 2007–2010, the Organic Law of Higher Education of 2010, la "Secretaría de Educación Superior, Ciencia, Tecnología e Innovación" (SENESCYT)'s public policy for the promotion of human talent, the decrees for the creation and restructuring of SENESCYT, SENESCYT's organic statute, SENESCYT ministerial agreements, among other official documents, are analyzed. This was complemented by the analysis of public interventions by main government authorities and bodies responsible for PS&T, especially those of the President Correa. Finally, we study some academic publications by government officials and academics critical of the reform showing the debate and the controversies present.

Additionally, 15 interviews were carried out with protagonists of PS&T definition and execution throughout the years: government authorities and public officials of the governing bodies of PS&T, former authorities of higher education governing bodies, current and former rectors and other authorities of the universities involved in these processes and researchers who benefit from PS&T.

Specifically, the documentary analysis and the interviews carried out seek to answer what characterized (dominant concepts, rationality, focus, financing, policy instruments, evaluation instruments, objectives, priorities) the Ecuadorian scientific and technological policy during these years. In addition, it is inquired who are the actors involved, how they have intervened, what relations or tensions have existed between them, or if there have been tensions between scientific development and technological development and why. It is also interesting to investigate how the state, through various macroeconomic, commercial and other policies, has promoted PS&T, and to what extent international models or paradigms have been assumed.

9.4 Results and Discussion

The renewed conceptions about PS&T in "*Buen vivir*" were expressed mainly in the National Development Plan 2007–2010 and in the Constitution of 2008. They contain novel and representative aspects of the PS&T proposal by the Correa administration in its early years: (1) the conception of government regarding the relationship between S&T and development, with emphasis is on social inclusion; (2) emphasis on the state's role in the promotion and coordination of S&T development; (3) the international policy on S&T; (4) the international conceptual frameworks in which the PS&T proposal is registered; (5) the concern for ancestral knowledge; and (6) the importance for the environment and biodiversity in PS&T.

Regarding the first, it was suggested that S&T should answer to economic and productive development and social objectives. The goal of social inclusion is a transversal element in PS&T and not just one more section in plans or programs. It is mentioned that S&T must strengthen national sovereignty and reduce dependency. President Correa said:

> we must reduce dependency on the large areas of the world that produce technology [...]. Our dependence on the centers of capitalism is accentuated to the extent that we seek in them technology which fascinates us, without at the same time seeking the scientific knowledge in its origin. The market reduced us to the condition of mere technology consumers, often behind schedule, and in deposits of technological waste. (Correa, 2010a)

The Development Plan relates to S&T with two goals: the development of human capacities and potentialities and the creation of a "solidary and sustainable" post-neoliberal economic system. All this while taking up discourses of the 1960s and 1970s, which understand S&T to respond to domestic problems, enhance local capacities and achieve national sovereignty, and boost economic growth, productivity and competitiveness (SENPLADES, 2007: 1 09–268). However, perspectives derived from the innovation economy are not abandoned. The notion of the national innovation system (Freeman, 1995; Lundvall, 1992; Nelson, 1993) is still present, which presumes that the generation of knowledge has the ultimate goal of generating innovation, that is, economically useful knowledge, and that innovation is a continuous and cumulative process through which companies strengthen their productive capacities based on the application of scientific and technological knowledge.

Second, it was suggested that promoting and fomenting S&T are the state's responsibility. The centrality that S&T takes in official discourse and in the public agenda should be highlighted; it expresses the political will by the government and translates into the decision to create permanent public resources. The Development Plan defines "scientific and technological development" as one of its macro-projects with a budget of $113.1 million between 2007 and 2010 ($28.3 million per year) (SENPLADES, 2007, p. 374). The Constitution establishes that the state must allocate the necessary resources for S&T (Constitution of Ecuador, 2008: article 388), and therefore defines budgetary pre-allocations for higher education and research, science, technology and innovation, among a few other sectors that enjoy this privilege (Constitution of Ecuador, 2008: article 298).

Third, in regard to international policy, the Development Plan proposed transferring technologies having direct social impact and promoting international cooperation, academic exchange and direct foreign investment, especially geared toward technology transfer and domestic capacity strengthening. The Constitution adds that it is the duty of the state to promote "research, scientific development and the exchange of knowledge and technology within Latin American integration" (Constitution of Ecuador, 2008: article 423 number 2). The latter aspires to create the South American Council of Science, Technology and Innovation of the "Unión de Naciones Suramericanas" (UNASUR), which according to Correa, was "fundamental for not having parallel agendas in research and wasting resources everywhere", and to possess an "evaluation mechanism, university accreditation and homologation of degrees" (Correa, 2009a).

Fourth, regarding normative frameworks or scientific technological paradigms, in the discursive definition of PS&T, three theoretical proposals are superimposed: science for the good of society, information society, knowledge society. First, the paradigm of "science for the good of society" (Vélho, 2011) because national styles are prioritized, local knowledge and cultural differences are assumed, and social welfare is focused on. Next, the Development Plan's goal is to incorporate the population into the "information society", a normative framework to conceive that information; and especially information and communication technology (ICT), the main tool for growth and development (Godin, 2009). It was proposed to democratize access to information, expand ICTs, electronic government, electronic commerce, online education and health, among other measures (National Development Plan 2007–2010, policy 11.6).

Finally, the Constitution says it is the state's responsibility "to facilitate and promote incorporation into society of knowledge to achieve the objectives of the development regime" (Constitution of Ecuador, 2008: article 387). Knowledge is understood as the main element to achieve the country's social and economic goals. Nevertheless, it should be mentioned that the dominant paradigm prevails, along with these three theoretical referents. That is the notion of national innovation system that emphasizes the idea of the need for knowledge production for technological and commercial use.

Fifth, the Development Plan and the Constitution formulated to recognize, value and protect ancestral knowledge (Constitution of Ecuador, 2008: article 57, numeral 12). Even, the Constitution went further. One of its major innovations was to define the system as "national system of science, technology, innovation and ancestral knowledge", a sui generis combination that in theory involves recovering, strengthening and enhancing ancestral knowledge to more than the classical purposes of creating, adapting and disseminating knowledge, and developing technologies and innovations to boost domestic production and increase efficiency and productivity (Constitution of Ecuador, 2008: article 385). This redefinition also meant expanding the typical conception of the actors that make up the system because in addition to the academy, the government and the private sector, natural or legal persons linked to ancestral knowledge were included (Constitution of Ecuador, 2008: article 386).

In accordance with Ecuador's being intercultural, the aim was to build bridges between Western science and the different wisdoms and ancestral knowledge. However, it surprises as a national innovation system, which justly raises a single rationality in the generation of knowledge: useful knowledge can coexist with the idea of promoting ancestral knowledge, which advocates the recognition of other rationalities that go beyond the idea of commercialization and commodification of knowledge. This is the particularity and ambiguity of the concept of "national system of science, technology, innovation and ancestral knowledge". Sixth, the Development Plan proposed research for the non-extractive use of biodiversity and its conservation, and to define a norm to protect traditional knowledge. It was thought that bio-knowledge would be the answer to the change in the accumulation model which the government will make explicit in later years.

Apart from these conceptual definitions, it is now interesting to analyze the conceptions and positions of the different participants in the reform of PS&T. It is important to understand how these relationships were forged since the beginning of the citizens' revolution, which necessarily implies referring to the reform of the higher education policy, closely related to the change in PS&T.

9.4.1 The Academy-Government Relations Regarding Higher Education Reform

The relationship between the "Citizens' Revolution" government and the academic community began in 2007 as a harmonious collaboration manifested in the charge that President Correa made to the "Consejo de Educación Superior" (CONESUP), the collegiate body comprised mainly of university rectors and in charge of higher education policy, of its most important project: the drafting the new constitution. Next, between 2008 and 2009, a respectful dialogue prevailed through round tables and academic events regarding university reform and the participatory construction of the new higher education law. However, as of 2010, with the approval of the "Ley Orgánica de Educación Superior" (LOES), the relationship between the government and at least a part of the university changed, characterized by a strong confrontation where divergences regarding reform of higher education policy and PS&T came to light. This relationship became increasingly complex. The increase in investment in higher education, S&T and other policies to promote these sectors have been well received by a good part of the academy. However, university reform and PS&T were the object of harsh criticism by another sector of the academic community.

In February 2007, President Correa entrusted CONESUP with drafting a preliminary bill that would serve as the basis for the new Constitution, based on the contributions of citizens. It seemed that relations between the government and the academy would be harmonious and that this and the political will of the government would influence a robust PS&T. Correa considered the Ecuadorian university to be "one of the few entities that enjoy prestige, truly representative of civil society, with moral and academic solvency" (Correa, 2007). This process showed coincidences between these two types of actors, since there were aspects of higher education reform that were on the academic community's agenda and that were consistent with the government's vision. It also showed divergences which, with other aspects, will mark future controversies in PS&T.

Three coincidences stand out: (1) improving the quality of higher education through evaluation and auditing, closing institutions that do not meet minimum standards and the prohibiting the creation of new universities until the constitutional reform is completed; (2) conceiving science, technology and innovation as a public good, focusing their development on solving social problems and therefore defining that it is the state's responsibility to promote them; and (3) creating a major in research. There were also discrepancies regarding freedom of research and the allocation of resources. CONESUP stated that research should have the "broadest freedom of creation" and that half of the state allocations should be distributed through CONESUP and the other half through competitive funds (CONESUP Constitution draft, articles 80–84). But, since mid-2007, the government delayed funds for S&T and centralized the evaluation of research and development (R&D) projects because they did not answer to national objectives. These caused discontent in university community who manifested their opposition through official documents to the Presidency and warnings of de facto measures.

During the debate on the new Constitution, it was decided to discuss problems of the higher education system and define agreements that would be taken to the Constituent Assembly. There were to be nine round tables for specific topics; the "Science, Technology and Education Forum in the New Constitution" and two "International Seminars for the Quality of Higher Education", one on evaluation and accreditation and another on admission and leveling. The agreements included: maintaining the budgetary pre-allocation for the university, defending autonomy, guaranteeing free higher education and establishing scholarships and educational credit to democratize access, creating an admissions system, linking permanence in the system to accreditation and quality evaluation, articulating the higher education system to national planning, encouraging basic research and technological development, establishing a major in teaching, among others (Rosales, 2015). However, other aspects came to light that would mark the discrepancy between government and the universities. University community aspired that the body responsible for PS&T be structured as a council that includes university representatives, researchers, society and government; and that the academic community participate in defining PS&T. These desires did not find an answer in the following years.

After the promulgation of the new Constitution, a higher education reform was initiated. It definitively reconfigured the relations between the government and the university community. Changes in PS&T understood almost as a subset of the higher education system were immersed, albeit with much less force. The urgency of university reform was shared between the academic community and governing authorities. Since the end of the 1980s, the university authorities had been expressing their concern regarding the lack of quality and relevance in their activities, evident in at least two investigations on the situation of the Ecuadorian university: "Evaluation of the current situation and perspectives for the short and medium term of the universities and polytechnic schools" (Pacheco, 1992) and "Mission of the Ecuadorian University for the Twenty-first Century" (CONUEP, 1994); and in its proposal to reform the Law of Universities and Polytechnic Schools in 1992. But this self-criticism did not result in concrete actions. The government understood university reform as a core and urgent element of its political project, manifested in the search for four transformations: (a) defining higher education as a "public social good", (b) large public investment, (c) quality improvement; and (d) the creation of a new institutionality (Correa, 2012a).

There was consensus in the first three proposals. However, the new law and new institutionality were the determining elements that broke the harmonious relationship between the government and the academy. The government disqualified the previous institutions responsible for higher education as well as the university actors who were organized and opposed to the proposal. It accused them of pursuing particular interests and proposed "breaking historical dynamics of cooptation of regulatory bodies by privileged groups" (Rosales, 2015, p. 103). Thus, it was able to "create and retain an interpretative monopoly on the understanding of the policy problem, which determined the dynamics of the problems and their solutions found" (Zabala, 2015, p. 47). This reform has been described as "winner take all" by Rosales (2015) quoting Linz (1993), or as "bureaucratic centralism" by Zabala (2015). In other words, there was a vertical control of decisions by the executive and little space for the participation or cooperation of the actors involved in this reform. On the contrary, through CONESUP authorities, the academic community stated that the process of defining the law was not legitimate because it was imposed by the government. They considered this law was punitive, regulatory and unconstitutional, that the dialogues were "only to serve and feign a space of participatory democracy", and that the proposals made were not collected (Vega, 2009).

In this confrontation, as a correlate of higher education reform, PS&T reform took place. Next, the characteristics of this policy change, the conceptions of the government and the academy regarding it, and the tensions between the actors will be presented.

9.4.2 Change of PS&T and Conceptions of the Relevant Participants

The LOES was not only the determining factor in the change in relations between the university and the government, but also the normative instrument that transformed PS&T. The law eliminated the "Secretaría Nacional de Ciencia y Tecnología" (SENECYT), the body responsible for PS&T (See Fig. 9.1), and joined higher education and S&T into a single state body: "Secretaría de Educación Superior, Ciencia, Tecnología e Innovación" (SENESCYT) (See Fig. 9.2). For the first time in Ecuador's history, the institutions responsible for PS&T had ministry status, but the collegiate body, where representatives of the academy and the private sector participated, disappeared. This change and the political will of the president and the large increase in public investment reflect the centrality that S&T has for this political project. However, for the government, PS&T is only understood in relation to, or even subordinated to, higher education policy.

The PS&T guidelines defined in the LOES focused on two aspects: the promotion of university research and the increase of critical mass through the financing of scholarships. As in other public sectors, this reform was made by making a difference with the past and under re-foundational rhetoric. The absence of PS&T and that few R&D resources were distributed according to the researchers' preferences (without planning and impact on development) was criticized. The importance that the government of the Citizens' Revolution gave to PS&T and the difference that

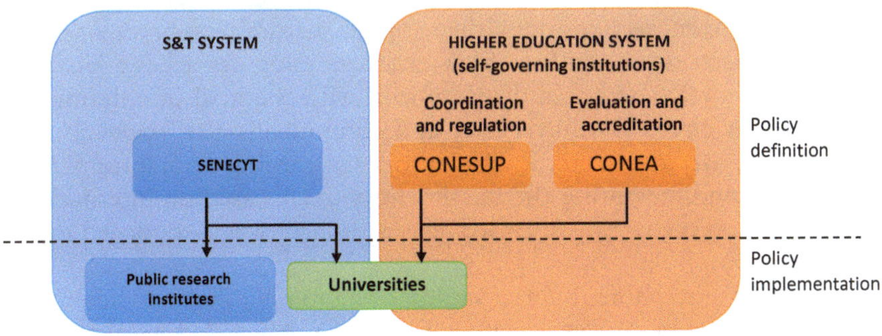

Fig. 9.1 Higher education system and science and technology system before 2010. CONESUP (Consejo Nacional de Educación Superior), CONEA (Consejo Nacional de Evaluación y Acreditación de la Educación Superior)

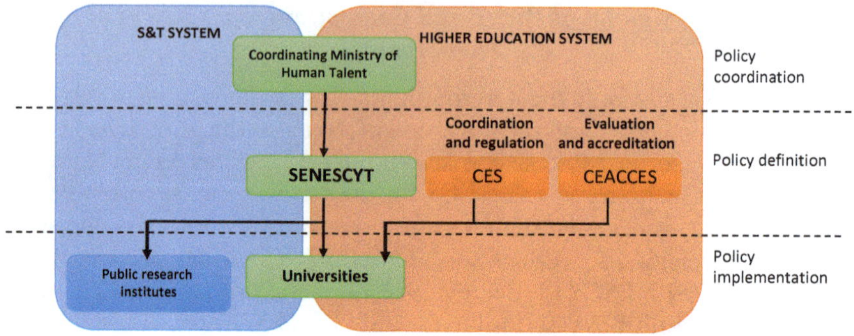

Fig. 9.2 Higher education system and science and technology system after 2010. CES (Consejo de Educación Superior), CEACCES (Consejo de Evaluación, Acreditación y Aseguramiento de la Calidad de la Educación Superior)

exists in relation to the past cannot be doubted. But in this article the important investment and specific instruments of PS&T are not analyzed. The analysis of the discrepancies between the national government and members of the academic community regarding the conceptual proposal of PS&T are of interest.

We begin by analyzing the beliefs of the leaders of the citizens' revolution on S&T under the premise that ideas explain policies because they intend to translate the interests of the actors into policies. The most important ideas that emerge from the public interventions by government authorities and the normative documents are: (1) human talent, knowledge and cultural change are the fundamental factors to achieve development, and are more important than capital, natural resources or other productive factors; (2) faith in S&T and technology as a basic factor to achieve economic and social development, with the goal of entering the knowledge society; (3) universality and communalism of science, paraphrasing the Mertonian ethos, which proposes learning from international experiences and measuring the fulfillment of goals based on international indicators; and recovering knowledge as a public good, as opposed to the patent system.

First, the government understands development from a heterodox perspective that privileges social welfare and the search for equity, and that's not limited to economic growth: "a welfare state" (Correa, 2008b). For the president, the necessary factors for development are (1) cultural; (2) human talent, understood as citizens with high academic studies;

(3) science and technology, or knowledge; (4) social, or having a national project; (5) institutional: the rules of the game that organize and order society; (6) physical capital; and (7) natural resources. He believes that "countries' keys to the success are no longer in these traditional production factors" (Correa, 2008a), but in the cultural heritage, human talent and knowledge (Correa, 2011c); he states that a cultural change based on achieving a culture of "excellence" and on ceasing to "fear" innovation (Correa, 2011d) is required. On the other hand, the importance given to human talent is evident in the policy of scholarships in the "best universities in the world", the PS&T instrument with the most resources during the first years, which Correa describes as "the most important thing that is making the revolution for the present and future of the country", in exchange for which, "new schools, hospitals, roads, hydroelectric power plants could be sacrificed, if necessary" (Correa, 2012b).

Second, we see a kind of "faith in science and technology" and a belief in the need and inevitability of entering the society of knowledge. They understand S&T, especially technology as an indispensable condition of "future" that will "do the same or more with less" (Correa, 2011d); as "cornerstone" of the economy and society (Correa, 2014b); as a way to eradicate poverty and other social problems (Correa, 2012b); as an element to modify social relations, productive relations and raise the level of society's welfare (Correa, 2011b); as "an instrument for individual freedom, social emancipation and to live and coexist well"; and as "a catalyst for economic and productive transformation" (SENPLADES, 2013, p. 160). Additionally, in re-foundational and libertarian rhetoric, Correa affirms that "a large part of our second and definitive independence is to become creators of knowledge and break that immoral international division of labor to which we want to submit" (Correa, 2014a). But, if we do not enter the society of knowledge, we are condemned to remain isolated, subordinated and dependent. Correa mentions:

> Today we know that in the knowledge society, which is contemporary society, what was seen yesterday as a dispensable luxury (which can only be enjoyed by those privileged with fortune or power), today is a need that cannot be postponed [...] because there is no possible horizon for those who remain lacking in knowledge, of the domain of the marvelous world of signs, whether for use in the rational representation of the world, as befits science, or imaginatively, as befits the arts. (Correa, 2009b)

In addition, in the official discourse we can find some closeness to technological determinism, specifically with certain approaches that emerged in the 1960s and 1970s regarding the "Theory of the Scientific and Technological Revolution" (Fleron, 1977; Richta et al., 1967). They proposed that both capitalist and communist countries would move toward a science-based production mode; or with authors proposing the theory of convergence (Bell, 1974; Galbraith, 1967), because they considered that all countries would move to post-industrial formation. Correa said:

> [...] I am convinced that scientific and technological advances can create much more well-being and be a greater engine for social changes than any class struggle or its other extreme, the simplistic search for individual profit. It would suffice to recall the industrial revolution, which transformed agrarian societies into industrial ones. (Correa, 2010b)

Third, there is a combination of notions around the universalism of science, tinged with the aspiration to produce one's own thought, and the idea that we are in a situation of lag or delay with regard to the advanced countries, so it is necessary to learn from them urgently. Universalism of science, as proposed by Merton (1942), considers that science is independent of national borders and local socio-cultural characteristics. Paradoxically, in Ecuador's case, thinking about S&T with one's own brains and under the principle of sovereignty is added to universalism. This is corroborated when the president says, "I am a critic of globalization, but in matters of excellence, I believe in globalization" and adds "we cannot close ourselves to global knowledge. We must cultivate ours as we learn from international experiences and knowledge, in a fine balance between openness, internationalist vision and our own sovereign thinking, to aspire to a higher education system that can be among the best in the world" (Correa, 2011c).

The characteristics of some of the applied policy instruments are justified over this vision, such as the allocation of scholarships at the "best universities in the world" or the Prometheus program to "learn from international experiences". International rankings and indicators are the reference for these policy instruments.

According to Merton (1942), communalism of science refers to the fact that "the substantive results of science are the product of social collaboration and are destined for the community", so that private appropriation of knowledge is excluded (Merton, 1942). Undoubtedly, this conception is present in that the government considers knowledge as "public good", "technically without capacity for exclusion and without rivalry in consumption" (Correa, 2010b), and in criticisms of the privatization of knowledge, copyrights and patents. Correa affirms that "the privatization of knowledge to maximize science and technology production" is a myth (Correa, 2010c); it creates social inefficiency and subjects people to the interests of capital because as there is no rivalry in consumption "the more people enjoy this already created good, the greater the social welfare". On the contrary, he affirms that there are "more efficient ways to encourage development of knowledge" such as state compensations for the creation of knowledge for profit or the greater participation of academia and the public sector" (Correa, 2010b).

Conversely, criticisms by academics, expressed in interviews, interventions in academic events and publications, are (1) the absence of an official PS&T proposal, the lack of a diagnosis of S&T reality, and the inexperience of SENESCYT authorities; (2) criticism of the universalist conception of S&T and the linear model of the development of S&T; and (3) the lack of relevance of PS&T, given they consider the proposal for the citizens' revolution does not depart from orthodoxy and that evaluation methods of scientific advances promote a kind of scientism (Varsavsky, 1969).

First, the government has not proposed a document defining PS&T, so that what has been carried out answers to changing perspectives, notions and interests of SENESCYT authorities expressed in ministerial agreements and lower hierarchy regulations. The lack of a diagnosis of this sector's reality is questioned and consequently, the improvisation with which investments are defined and the different policy instruments are applied. Villavicencio[1] (2014b) affirms that almost seven years have passed since the topics of science and technology, as axes of change in Ecuador's economy, were introduced in the government's agenda; however, there is no diagnosis, however elementary, that justifies the rationality of spending on scattered and doubtful actions (Villavicencio, 2014a, p. 3).

Cárdenas[2] (2013) considers that the lack of clarity in PS&T is due to "the inability to create a nucleus of people involved in S&T at the highest

[1] Joint Winner of the Nobel Peace Prize as part of the Intergovernmental Group of Experts on Climate Change (2007), President of CONEA (2009), Rector of the "Instituto de Altos Estudios Nacionales" (2012), Professor and researcher of the "Universidad Andina Simón Bolívar".

[2] Scholar of the program "Prometeo" in the "Escuela Politécnica Nacional" (2013–2014), University Professor (University of Texas and others).

level to lead this process" (Cárdenas, 2013, p. 2). He also mentions that it is strange that SENESCYT authorities are not required to have the same requirements as those to be university rector (LOES, article 49) or that they are not asked to have worked at a university. This begs the question, "If none of these people has worked at a 'university,' what enables them to be in a position of public service where university public policies are determined?" (Cárdenas et al., 2014, p. 35), which clearly shows a confrontation between representatives of "academic culture" and the "political culture".

Second, for some academics, the Ecuadorian university reform must be understood as part of international rationalization similar to others in the region and the Bologna Process itself. Carvajal[3] (2013) considers that the "illusions that underlie" this rationalization tries to put the universities at the top of the rankings, believe there is a necessary and immediate relationship between the quality of universities and social and economic development indices, and think that with more PhD's we will compete in areas of the technological frontier (Carvajal, 2013, p. 60).

For Villavicencio (2013) this universalist conception, besides expressing itself in the internationalization of higher education and in the use of typologies, rankings and international indicators, translates into a commitment to "great science", revealed more clearly in the proposal of the city of knowledge "Yachay" "designed on the model of two U.S. universities, [and], moreover, all the scientific-technological complex or the so-called city of knowledge, planned by a South Korean company" (Villavicencio, 2013, p. 219). The government's proposal is based on the linear model technology push, overcome in other latitudes decades ago (Espinosa 2015,[4] Villavicencio 2013).

Third, several authors (León, 2013; Muñoz, 2013) say the reform of PS&T and higher education in Ecuador and other post-neoliberal countries would be closer to a new process of industrialization that an alternative to development. According to León (2013), the Ecuadorian reform is similar to the World Bank's proposals expressed in the book "Closing the

[3] General Secretary of the "Consejo Nacional de Universidades y Escuelas Politécnicas" (CONUEP) (1987–1991), Director of the Mission Project of the Ecuadorian University for the twenty-first century, Director of Research at the "Pontificia Universidad Católica del Ecuador" (2005–2010).

[4] Alfonso Espinosa (Rector of the "Escuela Politécnica Nacional" (1990–1994, 2009–2013), former President of CONUEP's Planning Commission), in conversation with the author.

gap in education and technology" (World Bank, 2005) because both say that since the 1960s productivity and growth in the region have been stagnant, unlike what happened with the "Asian tigers". Also, because the two assume these gaps are due to differences in capabilities and technology. The World Bank states that "the key to development would be creating national systems of innovation, importing technology and educating workers in skills, with state collaboration ('the government') with private enterprise"(World Bank, 2005, p. 6), while the:

> Ecuadorian government [...] advocates that 'the keys to development, to buen vivir, are more than the quantity of human work, human talent, the quality of that work and science and technology' (Correa, 2011a, p. 6), an achievable goal providing state support to the private sector. (León, 2013, p. 89)

Then, apparently, what is of interest is all "useful knowledge" (Hurtado, 2015). Even from some more radical voices it is argued that the reforms of PS&T and higher education are just another expression of a development model that cannot overcome neoliberalism. Muñoz (2013) says this:

> model of capital accumulation and/or productive matrix [...] reorients its insertion in the globalized world (especially BRICS countries, particularly China), promoting and disputing capital and markets, as well as the development of science and technology. (Muñoz, 2013, p. 123)

Similarly, Acosta (2013) considers this is the result of a "regional historical change determined by the crisis of global hegemony, which leads in this new historical stage to the return of the state and post-neoliberalism" and that the state solves local problems and creates an image of change and modernization, for example, the first Ecuadorian satellite placed in orbit in April 2013 (Acosta, 2013, p. 15).

9.5 Conclusions

The definitions of PS&T included in the National Development Plan 2007–2010 and the Constitution of 2008 show some novel elements that move away from hegemonic conceptions about development. At the same time, these definitions of PS&T preserve and value certain elements of development orthodoxy and the dominant conceptions of PS&T. Just like other post-neoliberal processes, the citizens' revolution in general, and PS&T reform in particular, presents themselves as spaces of dispute in which advances appear that go beyond the neoliberal proposal, but where many of their elements are maintained.

Among the advances are (1) social inclusion and economic development as a goal of scientific and technological development; (2) the importance of S&T, emphasizing the state's role in financing and promoting S&T and surpassing the reality of the 1990s and previous decades, where the Ecuador did not contemplate allocation of considerable or permanent resources for S&T; and (3) the recognition, assessment, recovery and inclusion of ancestral knowledge in the national system of science and technology, in building an intercultural state. On the other hand, elements inherited from the neoliberal era are (1) the prioritization of economic growth and the improvement of productivity and competitiveness as central objectives of S&T; and (2) the uncritical preservation of conceptual frameworks of the mainstream, included as referents and models in building PS&T: information society, knowledge society and national innovation system.

However, later normative definitions of PS&T, specifically LOES, and the official discourses that guided the application of PS&T are the elements around which discrepancies between the academic community and the government are evident. After a period of coordination and dialogue between these two actors, these differences are born with the same promulgation of LOES and with the institutional reform of higher education politics and PS&T governing bodies. The new institutional structure centralizes these policies in the government and eliminates the historical control the academic community had over them. There are divergences around several conceptual elements of the new PS&T proposal. Critical academics call attention to aspects of PS&T that question the same project of the citizens' revolution, wondering if this exceeds the orthodoxy of development or if it only rearranges itself in capitalist reordering. They denounce the use of international indicators and rankings, which they consider the expression of erroneous models and misconceptions related to the ideas of the universalism of science and the linear model of S&T development, and consequence of pressure for higher education institutions to participate in international classifications such as Times Higher Education World University Rankings or QS World University Rankings. In the substantive debate, the latent question is whether this change in PS&T is a new uncritical adoption of international models or if in reality post-neoliberal rhetoric is the guideline for PS&T. This translates into a discussion about some elements of the most advanced Western economies: the private appropriation of scientific knowledge, the commercialization and commercialization of research, the transfer of technology, the delimitation of international borders and patenting, among others. However, it remains to be seen how these contradictions are presented in the implementation of PS&T and its instruments. Undoubtedly, it will be possible to better observe the real advances of PS&T and its distances with the orthodoxy of development.

References

Acosta, A. (2013). A modo de prólogo. In J., Cuvi, et al. (Eds.), *El correísmo-Un nuevo modelo de dominación burguesa* (pp. 9–21). Quito.

Averch, H. (1985). *A Strategic Analysis of Science and Technology Policy*. Baltimore: The John Hopkins University Press.

Banco Mundial. (2005). *Cerrar la brecha en educación y tecnología*. Bogotá: Banco Mundial-Alfaomega Colombiana.

Becker, H. (2009). *Trucos del oficio: cómo conducir su investigación en ciencias sociales*. Buenos Aires: Siglo XXI Editores.

Bell, D. (1974). *Coming of Post-industrial Society*. Nueva York: Basic Books.

Blume, S. (1985). *The Development of Dutch Science Policy in International Perspective, 1965–1985*. Report to the Raad van Adviesvoor het Werenschapsbeleid (RA WB), Study no 14.

Bonvillian, W. (2011). The Problem of Political Design in Federal Innovation Organization. In Husbands, et al. (Eds.), *The Science of Science Policy A Handbook*. Stanford University Press.

Brooks, H. (1982). Science Indicators and Science Priorities. In M. La Follette (Ed.), *Quality in Science*. Cambridge: The MIT Press.

Cárdenas, J. (2013). *Plan Preliminar Para Estimular las Ciencias Básicas Como Elemento Crítico del Desarrollo de la Ciencia y Tecnología en el Ecuador*. Quito: Escuela Politécnica Nacional.

Cárdenas, J., et al. (2014). *Lo Común de las Virtudes: De Banana Republic a Banana Republic Tecnológica*. Quito: Escuela Politécnica Nacional.

Carvajal, I. (2013). *Tecnocracia y democracia en el ocaso de la "Universidad Ecuatoriana"*. Ecuador Debate, 90, diciembre, 47–64.

Consejo Nacional de Universidades y Escuelas Politécnicas (CONUEP). (1994). *Universidad Ecuatoriana. Misión de la para el siglo XXI. Conclusiones y Propuestas*. CONUEP- EB/PRODEC, Quito.

Constitution of Ecuador. (2008). http://www.hlrn.org/img/documents/Constitucion_del_Ecuador_2008.pdf

Correa, R. (2007). *Discurso del presidente Rafael Correa al presentar la Comisión del CONESUP para codificar la propuesta para la Asamblea Constituyente*. Discurso 28 de febrero, Quito.

Correa, R. (2008a). *Socialismo del siglo XXI*. Conferencia, 11 de abril, México DF.

Correa, R. (2008b). *Intervención del presidente de la República, Rafael Correa en el cierre de la campaña por el sí*. Discurso, 24 de septiembre, Guayaquil.

Correa, R. (2009a). *Ceremonia de traspaso de la presidencia pro tempore de la UNAS U R*. Discurso, 1 O de agosto, Quito.

Correa, R. (2009b). *Premios "Eugenio Espejo"*. Discurso, 20 de agosto, Quito.

Correa, R. (2010a). *Universidad Central del Ecuador*. Discurso, 18 de marzo, Quito.

Correa, R. (2010b). *Seminario Científico "Antártica, un Nuevo Norte para Magallanes"*. Conferencia, 5 de noviembre, Chile.

Correa, R. (2010c). *Conferencia magistral del economista Rafael Correa Delgado en la Universidad de Buenos Aires*. Conferencia, 03 de diciembre, Buenos Aires.

Correa, R. (2011a). *Prometeo "Viejos Sabios"*. Discurso, 9 de febrero, Quito.

Correa, R. (2011b). *Creación del Ministerio Coordinador de Conocimiento y Talento Humano. Posesión de Ministro*. Discurso, 11 de abril, Quito.

Correa, R. (2011c). *Conferencia en el lII Congreso Internacional Universidad, Desarrollo, y Cooperación*. Conferencia, 10 de mayo, Quito.

Correa, R. (2011d). *Evento de premiación a la excelencia académica*. Discurso, 18 de octubre, Guayaquil.

Correa, R. (2012a). *Ciencia y Fe. 50 Años Universidad Católica Santiago de Guayaquil*. Conferencia, 17 de mayo, Guayaquil.

Correa, R. (2012b). *Intervención en la adjudicación de becas – convocatoria abierta 2012*. Discurso, 29 de agosto, Quito.

Correa, R. (2014a). *Inauguración de los cursos de nivelación y del campus patrimonial de Yachay "ciudad del conocimiento"*. Discurso, 31 de marzo, Urcuquí.

Correa, R. (2014b). *Ecuador y sus transformaciones en política, ciencia y tecnología*. Conferencia en la Universidad de Yale, 10 de abril, New Haven.

Dickson, D. (1984). *The New Politics of Science*. Chicago: The University of Chicago Press.

Elzinga, A. (1988). From Criticism to Evaluation. In A., Jamison (Ed.), *Keeping Science Straight. A Critical Look at the Assessment of Science and Technology*. Proceedings of a Conference in Gothenburg, June 1987. Report no 156, 10 May, pp. 29–58.

Elzinga, A., & Jamison, A. (1996). El cambio de las agendas políticas en ciencia y tecnología. *Revista Zona Abierta*, 75/76, Madrid.

Fleron, F. (1977). *Technology and Communist Culture: The Socio-cultural Impact of Technology Under Socialism*. Nueva York: Praeger.

Freeman, C. (1987). Quantitative and Qualitative Factors in National Policies for Science and Technology. In J. Annerstedt (Ed.), *From Research Policy to Social Intelligence. Essays in Honour of Stevan Dedijer*. London: Macmillan.

Freeman, C. (1995). The 'National System of Innovation' in Historical Perspective. *Cambridge Journal of Economics, 19*(1), 5–24.

Galbraith, J. (1967). *The New Industrial State*. Londres: Hamish Hamilton.

Geddes, B. (2006). *Paradigms and Sand Castles: Theory Building and Research Design in Comparative Politics*. Michigan: University of Michigan Press.

Godin, B. (2009). *The Making of Science, Technology and Innovation Policy: Conceptual Frameworks as Narratives, 1945–2005*. Montréal (Québec): Centre Urbanisation Culture Société Institut national of the recherche scientifique.

Huanacuni, F. (2010). *Buen Vivir/ Vivir Bien. Filosofía, políticas, estrategias y experiencias regionales andinas*. Coordinadora Andina de Organizaciones Indígenas – CAOI

Hurtado, D. (2015). La cultura científico-tecnológica argentina en contexto democrático: tres etapas. In A. Ferrer et al. (Eds.), *Universidad pública y desarrollo: innovación, inclusión y democratización del conocimiento* (pp. 34–47). Buenos Aires: IEC – CONADU/CLACSO.

León, C. (2013). *Cambiar el alma ... ¿exiliar a las humanidades?*. Ecuador Debate, 90, diciembre, 87–102.

Levi, M. (2006). Modelando procesos históricos complejos con narrativas analíticas. *Revista Uruguaya de Ciencia Política, 15*(1), Instituto de Ciencia Política, Uruguay, 11–29.

Linz, J. (1993). *Los peligros del presidencialismo, Comisión Andina de Juristas: Formas de gobierno: relaciones Ejecutivo-Parlamento*, Lima.

Lundvall, B. (1992). *National Systems of Innovation, Towards a Theory of Innovation and Interactive Learning*. London: Frances Pinter.

Marí, M. (1982). *Evolución de las concepciones de política y planificación científica y tecnológica*. Temas de Política Científica y Tecnológica. Documento del trabajo No. 1, OEA, Washington, DC.

Merton, R. (1942). La estructura normativa de la ciencia. In R. Merton (Ed.), *La Sociología de la Ciencia* (pp. 355–368). Madrid: Alianza.

Muller, P. (2009). *Las políticas públicas*. Colombia: Universidad de Externado.

Muñoz, F. (2013). Forma de Estado y régimen político en el gobierno de Rafael Correa. In J. Cuvi et al. (Eds.), *El correísmo al desnudo* (pp. 120–133). Quito: Montecristi Vive.

Nelson, R. (1993). *National Innovation Systems: A Comparative Analysis*. New York: Oxford University Press.

Pacheco, L. (1992). *Resumen del Informe: Evaluación de la Situación actual y perspectivas para el corto y mediano plazos de las Universidades y Escuelas Politécnicas*. Quito: CONUEP.

Palacios, D. (2014). *Sumak kawsay, hacia la política pública del buen vivir*. Tesis de Maestría, FLACSO, Ecuador.

Postolski, G., & Funes, P. (2015). Prefacio. In A. Ferrer et al. (Eds.), *Universidad pública y desarrollo: innovación, inclusión y democratización del conocimiento* (pp. 10–12). Buenos Aires: IEC-CONADU/CLACSO.

Powell, W., Owen-Smith, J., & Smith-Doerr, L. (2011). Sociology and the Science of Science Policy. In Husbands, et al. (Eds.), *The Science of Science Policy A Handbook* (pp. 31–55). Stanford University Press.

Ramírez, R. (2010). La transición ecuatoriana hacia el Buen Vivir. In I. León (Ed.), *Sumak Kawsay/Buen Vivir y cambios civilizatorios* (pp. 125–142). Quito: FEDAEPS.

Ramírez, F. (2013). *El proceso constituyente ecuatoriano y sus dilemas de legitimidad. Un contrapunto andino. Perspectivas constitucionales comparadas.* Bologna: Universidad de Bologna (en prensa).

Richta, R., et al. (1967). *Civilization at the Crossroads: Social and Human Implications of the Scientific and Technological Revolution.* White Plains, NY: International Arts and Sciences Press.

Rip, A., & Hagendijk, R. (1988). *Implementation of Science Policy Priorities.* Science Policy Support Group (SPSG), Concept Paper no 2, London.

Rosales, R. (2015). *El sistema político ecuatoriano y su capacidad de generar políticas públicas: el caso de la ley orgánica de educación superior.* Tesis de maestría, FLACSO, Ecuador.

Rothwell, R., & Zegveld, W. (1985). *Reindustrialisation and Technology.* London: Longman.

Ruivo, B. (1994). Phases and Paradigms of Science Policy. Science and Public Policy, 21, num 3, 157–163.

Salomon, J. J. (1977). Science Policy Studies and the Development of Science Policy. In I. Spiegel-Rosing & D. Price (Eds.), *Science, Technology and Society: A Cross-disciplinary Perspective* (pp. 43–70). Londres: Sage.

Sanz, L. (1997). *Estado, ciencia y tecnología en España: 1939–1997.* Madrid: Alianza Editorial.

Sanz, L. (2001). ¿Por qué cambian las políticas? La política europea de investigación y desarrollo tecnológico. *Revista Española de Ciencia Política, 4,* 97–121.

Scartascini, C., Spiller, P., Stein, E., & Tommasi, M. (Eds.). (2011). El juego político en América Latina: ¿Cómo se deciden las políticas públicas?. BID.

Secretaría Nacional de Planificación y Desarrollo (SENPLADES). (2007). *Plan Nacional de Desarrollo 2007–2010.* Quito: SENPLADES.

SENPLADES. (2013). *Plan Nacional de Desarrollo/ Plan Nacional para el Buen Vivir 2013–2017.* Quito: SENPLADES.

Stoessel, S. (2014). Giro a la izquierda en la América Latina del siglo XXI. Revisitando los debates académicos. *Polis. Revista Latinoamericana, 13*(39), 123–149.

Streeck, W., & Thelen, K. (2005). Introduction: Institutional Change in Advanced Political Economies. In W. Streeck & K. Thelen (Eds.), *Beyond Continuity: Institutional Change in Advanced Political Economies* (pp. 1–39). Oxford: Oxford University Press.

Varsavsky, O. (1969). Ciencia, política y cientificismo. Centro Editor de América Latina, Buenos Aires.

Vega, G. (2009). Oficio No. 895, CONESUP.PR del presidente del CONESUP dirigido a Alberto Acosta, presidente de la Asamblea Constituyente, 14 de septiembre.

Vélho, L. (2011). La ciencia y los paradigmas de la política científica, tecnológica y de innovación. In A. Arellano & P. Kreimer (Eds.), *Estudio social de la ciencia y la tecnología desde América Latina* (pp. 99–125). Bogotá: Siglo del Hombre Editores.

Villavicencio, A. (2013). ¿Hacia dónde va el proyecto universitario de la Revolución Ciudadana? In J. Cuvi et al. (Eds.), *El correísmo al desnudo* (pp. 216–231). Quito: Montecristi Vive.

Villavicencio, A. (2014a). Universidad, conocimiento y economía. Pre-textos para el debate. N° 1. Universidad Andina Simón Bolívar, Quito.

Villavicencio, A. (2014b, May 2). La fiebre de las publicaciones. La línea de fuego. https://lalineadefuego.info/2014/05/02/la-fiebre-de-las-publicaciones-por-arturovillavicencio/

Whitley, R. (2010). Reconfiguring the Public Sciences: The Impact of Governance Changes on Authority and Innovation in Public Science Systems. In R. Whitley, J. Glaeser, & L. Engwall (Eds.), *Reconfiguring Knowledge Production: Changing Authority Relationships in the Sciences and their Consequences for Intellectual Innovation* (pp. 3–47). Oxford: Oxford University Press.

Zabala, T. (2015). *Análisis del cambio de la política pública de educación superior en Ecuador, período 2008–2013*. Tesis de maestría, FLACSO, Ecuador.

CHAPTER 10

Against All Odds: Scientific Education, Scientific Competences and the Ideals of an Educated Country, Colombia 1995–2010

Gina G. Calderón and Edgar G. Eslava

10.1 Introduction

As evidenced by the use of standardized assessment systems and national sets of policies for the basic, intermediate and higher education levels, there seems to be a cross-border agreement on the type of orientation that should be given to the scientific education programs (OECD, 2007; Winterton, Delamare-Le Deist, & Stringfellow, 2006). The underlying view is the one in which education is the natural space for the development

G. G. Calderón (✉)
Oficina de Innovación Educativa, Ministerio de Educación Nacional, Bogotá, Colombia
e-mail: gcalderon@mineducacion.gov.co

E. G. Eslava
Facultad de Filosofía y Letras, Universidad Santo Tomás, Bogotá, Colombia
e-mail: edgareslava@usantotomas.edu.co

C. I. Pardo Martínez et al. (eds.), *Analysis of Science, Technology, and Innovation in Emerging Economies*,
https://doi.org/10.1007/978-3-030-13578-2_10

of skills imperative for citizens to reach the minimum levels of dignified and responsible life in their personal and social context, in line with the images of professionalization and citizenship that will prevail and will be needed to consolidate social, political and economic projects of the nations (Rychen & Salganik, 2003; UNESCO, 2005a).

The agreement is, or seems to be, that it is necessary to train in skills, ways of intervening the world based on a deep understanding of one's own context, the local problems needed to be resolved and the personal capabilities and expectations for intervening and resolving. Accordingly, it would be important that we move toward an education that promotes the development of competences that are considered indispensable for the development of individuals and societies and leave behind an education centered solely on conceptual contents from different areas of knowledge.

But the reach and scope of a skills-based education is far from clear and explicit for all who play a role in the design and implementation of national education policies. Changing traditional—concept-based—approaches to teaching and learning is not easy, leaving educators and policy makers with a pressing task, the "daunting challenge of creating new discourses, pedagogical practices, and collective strategies that will offer students and others the hope and tools necessary to revive the culture of politics as an ethical response to the demise of democratic public life" (Giroux & Searls, 2004).

While the spirit and the interest in an education based on actions, contexts and realities, the epistemological core of competence training is a mostly shared vision. There is no clear and consensual definition of what a competence is, what it means to be competent, and even less about the theoretical discourse of competences or the impacts and implications they have for the definition of future social practices and policies. In a nutshell, there is still no clarity about what it means to implement a new educational perspective centered on the development of competences (Carabaña, 2011; Figazzolo, 2009). If the above is a general problem for an education system that bases its policies and strategies on the mentioned development, the situation is no different within the specific context of science education, even more in regions of the world where the "knowledge economy" is still a vague concept, even a utopia.

In the specific case of Colombia, a country with an economy mostly based on agricultural production and marked by extreme inequalities in terms of access to basic resources, among them to a quality education, the promise of a scientific education that allows students to gain access to theories and practices and offers them opportunities for an "integral, equitable and sustainable human development" (Ministerio de Educación Nacional, 1998) is still to be fulfilled. Such an education presupposes a meaningful learning of the sciences, that includes the understanding of conceptual categories, the solution of problems and the development of basic competences that would allow students to become successful citizens able to participate in fields as diverse as academic tertiary programs, manual and technical work, and family and community life. Under such a view, educating in science is more than teaching theories and experiments to a group of young students in a classroom, it is helping citizens to understand the physical world, the relationships between its parts and the connections that exist between scientific work and other social activities; to participate in the construction of a society that knows the dynamics of scientific activity (Eslava, 2014).

Within this general framework, in which science has such a fundamental social role, it is necessary to educate citizens aware of the technologies that permeate their life, capable of creating and transforming knowledge to answer the problems whose solutions are important for their communities. But this task is a difficult one because, just as there is no clarity in the general discourse of competences—nor at the level of scientific competences—there is also no clear path to incorporate them into the current pedagogical practices or the present-day educational policies.

These are some of the issues framing the present research. Our work addresses the meanings, relations, implementation and evaluation of scientific competences that are present in the official Colombian educational policy, in order to understand the way in which such competences are to guide the transformation of the country into a scientific educated one.

10.2 Literature Review

During the course of the research, we found recurrent references to international agencies and projects that have played a central role in the definition of competences, their uses and scope, as part of Colombian educational system. To those organizations and projects, we turn our attention in this moment, focusing in the elements that are going to affect in the configuration of the country's scientific education policies. The presentation will advance from the more general and inclusive agendas for the definition and insertion of scientific competences—those of UNESCO and its affiliated organizations—to the ones that attempt to use those competences in explicit geographical and political contexts, and from there, to specific projects designed to measure the results of the inclusion of scientific competences in national and regional educational programs. The section concludes with an overview of the studies about scientific competences in Colombian education policies.

10.2.1 UNESCO on Scientific Education and Competences

Participants in the 1999 World Conference on Science for the Twenty-first Century, gathered in Budapest, organized by the United Nations Educational, Scientific and Cultural Organization (UNESCO) and the International Council for Science (ICSU), presented a series of directives that would serve as a framework to address the problems of scientific research, respond to the challenges it poses and promote collaboration among all the protagonists of scientific work, both nationally and internationally.

The agreements reached in the conference were considered as an engine that powered some of the strategies and policies that UNESCO has promoted for education in science and technology, from the very early initial proposals (Faure et al., 1973; UNESCO, 1966, 1997) until very recent recommendations (Care & Luo, 2016; Halinen, 2017; Roegiers, 2016; UNESCO, 2007a, 2007b, 2011a, 2015), emphasizing the need to integrate scientific and technological knowledge into the development of a culture that promotes the inclusion, both of the results and of the tools of the sciences, in increasingly complex social contexts, and in which "everything falls into place, if it is the requirements of science and technology, knowledge of self and of the environment, or the development of skills enabling each person to function effectively in a family, as a citizen or as a productive member of society" (Delors et al., 1996, 13).

According to *The Treasure within* report, regarding scientific education:

> It is necessary, among other things, to become familiar at an early age with science and the uses of science, and with the difficult task of assimilating progress in such a way that human identity and integrity are fully respected. Here, too, the ethical issues must not be overlooked. It also shows that the Commission is aware of the contribution that education must make to economic and social development. The education system is all too often blamed for unemployment. This observation is only partly true; above all it should not obscure the other political, economic and social prerequisites for achieving full employment or enabling the economies of underdeveloped countries to take off. (Delors et al., 1996, 20)

With these results in mind, the participants in the Budapest conference—where the teaching of the sciences was a central concern for all stakeholders—agreed on a number of issues that were to mark further approaches to the construction of educative policies in the entire world. Those agreements included considerations about the roles of governments, educative institutions, non-governmental organizations, pre-service and in-service teacher-training centers, as well as national, regional and international authorities, for all of whom a number of commitments were defined. Among those:

- Governments should give the highest priority to the improvement of science education at all levels, with special attention to eliminating the effects of gender disparity and discrimination against marginalized groups, as well as public awareness and the vulgarization of science. Steps must be taken to professionally prepare teachers and educators to cope with the changes, and also to do everything possible to combat the shortage of teachers and educators who are properly qualified to teach science.
- Science teachers of all levels and personnel who take part in non-formal scientific education should be able to permanently update their knowledge, in order to be able to carry out their pedagogical work in the best possible way.
- In order to respond to the changes that occur in the educational needs of our societies, national education systems must renew curricula, methodologies and resources, taking into account the equality between the sexes and cultural diversity. Research on scientific and technical education will be promoted at the national and international levels, creating specialized centers and fostering the interconnection of those already existing throughout the world, with the cooperation of UNESCO and other competent international organizations.
- Educational establishments must provide a basic scientific education to students who do not pursue a scientific career. They must also provide opportunities for ongoing scientific training. (UNESCO, 2000).

The scope of these commitments with the scientific training was later extended, within the timeframe that interests us here, to questions about the quality of education (UNESCO, 2002, 2004a), inclusion and cultural diversity (Hirmas, 2009), educational practices (UNESCO, 2004b), the so-called new humanism (UNESCO, 2011b), participation in the knowledge of societies (UNESCO, 2005a), and the key competences in education (UNESCO, 2005b). All of these making use of an idea of "educational justice" that, of course, would include science education as an integral part.

At a more local level, in Latin America, those international agendas were followed by a series of meetings and their respective declarations: Sao Paulo declaration (UNESCO, 2000), Buenos Aires declaration (UNESCO, 2004c), Lima declaration (UNESCO, 2014), in which the governments of the regions committed to the promotion of educational systems that guarantee quality education, universal coverage, and which specific budgetary assignments, from each country, must be made to ensure the incorporation of new technologies in the educational systems and practices (Eslava, 2015).

As an active UNESCO member, Colombia has subscribed to all the major education agreements and declarations, committing will, resources and individuals to the fulfillment of the goals established by the educational commissions. Among those are the eight Millennium Development Goals (MDGs) and the Education for All goals (EFA). Both MDGs and EFA are now updated as part of the so-called post-2015 agenda, included into the current Sustainable Development Goals (SDGs). As such, there is an explicit adherence to the competences-based models promoted by UNESCO and therefore an explicit use of its conceptual framework as an integral part of the country's educational policies.

10.2.2 European Union "Key Competences"

In November 2005, the Commission of the European Communities submitted for its consideration the recommendation "on key competences for lifelong learning". This recommendation was then adopted by the European Parliament and the Council at the meeting of December 18, 2006 (European Parliament, 2006):

> [...] defines the key competences that all citizens need in knowledge-based economies and societies. Recognizes the desirability of implementing decisions taken at the national, regional and local levels. Urges the Member States to ensure that all people have acquired the key competences at the end of initial education and training and, in light of European reference levels, encourages them to address the imbalances in education.

As part of the explanatory memorandum of the recommendation, the Commission traces its history, showing why, from the establishment of "adaptation to globalization and the shift towards knowledge-based economies" as a common task, all the countries of the union, later sanctioned by the Council of Europe in the year 2000 (Consejo de Europa, 2000), agreed on the importance of guaranteeing all citizens the possibility of accessing an education that offers them the knowledge and skills to participate in an economy that is increasingly focused on the development and communication processes of information. Such an agreement made it, in turn, essential to count with a common frame of reference that would allow the designing, implementing and evaluating of educational policies and practices tending to the acquisition of such competences.

A framework like that would complete the work on fundamental competences previously advanced by the working group on basic competences that the Council, within the "Education and Training 2010" program, established by the beginning of the new century, and according to which elementary mathematics and the ability to read and write in the mother tongue are the minimum necessary for a citizen to perform satisfactorily in society. To these skills were later added those of development of entrepreneurship and the competences competition in Information Technology and Communication (ICT) (Consejo de Europa, 2002).

According to the Commission, when defining the key competences, and in order to make explicit the fact that they extend the notion of "basic qualifications" in force since 2000, "the working group preferred to use the term 'competence' to refer to a combination of knowledge, skills and attitudes, and 'key competence', to define the necessary competences for all this. In this way, the concept of basic capabilities is included, but it is not limited to them" (Comisión de las Comunidades Europeas, 2005, p. 2).

As a result, the Parliament final recommendation defines competences as "the combination of knowledge, skills and attitudes appropriate to the context" (European Parliament, 2006, p. 12) and establishes as key ones communication in a foreign language, mathematical competence, basic competence in science and technology, digital competence, learn to learn, social and civic competences, sense of initiative and entrepreneurship, consciousness and cultural expression. These "key competences", developed to contribute to the consolidation of a knowledge society and to be interpreted as part of the essential foundations for learning, have demonstrated to be valuable for reaching agreements at the interior of the European commonwealth, and even beyond its frontiers—as it is demonstrated by the inclusion of some of its definitions, its educational and evaluative strategies and even the justification for its use and promotion at the very heart of some of the documents we found during the present study.

10.2.3 Competences in OECD DeSeCo Project

The Definition and Selection of Competences: Theoretical and Conceptual Foundations Program, DeSeCo, of the Organization for Economic Cooperation and Development (OECD), was a project that run from 1997 to 2002 under the auspice and management of the Swiss Federal Statistical Office, whose main goal was to develop "a coherent strategy for defining, selecting and measuring competences and skills" (OECD, 2001). Not being the actual definition of the key competences its core objective, DeSeCo advanced a framework from which later in time such definitions could be made. However, the program is relevant because it set some of the language, methods and horizon adopted by a number of agencies (UNESCO, OECD, World Bank, among others), countries and policy-makers in their efforts for establishing their own educational projects, programs and agendas, as is the case of the European Parliament "key competences" definition task we presented in the previous section.

In its characterization of competencies, DeSeCo starts from the question about how individuals and society can adapt to each other: What competences does a person need to achieve a successful life? How would society work better? From this reflection it is concluded that the number of competences that an individual must develop in order to answer these questions must be small enough for its presence, development and effectivity to be able to be measured (Stiefel, 2008).

The program suggests the adoption of a competence as "the ability to meet individual or social demands successfully, or to carry out an activity or task", defending what is calls a functional, or demand oriented, approach as one that gives a relevant place to "the personal and social demands facing individuals" (OECD, 2002, p. 8), taking a stance about the actual need of individual for adapting to previously unencountered situations, which, in turn, will translate to the educative context as the need for making use of a number of different strategies when dealing with new knowledge, or the application of the old one to a new context.

Accordingly, the program highlights the importance of not paying attention only to the cognitive skills and basic knowledge, but to include other aspects such as values and motivation, given that each competence is built on a combination of cognitive skills and interrelated practices. Knowledge (explicit and tacit) motivation, values, emotions and other social and behavioral components are mobilized at the same time to carry out effective action. Conceptualized this way, competences are observed in the actions carried out by individuals in specific situations or contexts, making DeSeCo's functional definition a holistic approach that integrates the external demand with the internal capacities of the individual.

Once what a competence is defined, and with the objective of facing current challenges in education in mind, DeSeCo establishes three categories of competences and proposes the scope of each of the defined categories as follows (OECD, 2003). The first category integrates competences related to the interactive use of tools. Here, a tool is seen not as a passive mediator, but as an instrument in dialogue between the individual and his environment. The type of competence included in this category requires familiarity with the tool and an understanding of how to interact with the world when using the tool. In this category, we find the interactive use of language, symbols and texts, knowledge and information, and technology.

The second group of key competences includes the consideration of increasing social diversity in Western societies, to achieve the benefit of individuals and to improve levels of cooperation in different cultures. This category has to do with what people need to learn, live and work with one another, and includes following competences: relate well with others, cooperate with others, and address and solve problems.

The final group is related to the training of people who can develop self-control of their own life to function properly in different social, family and work environments. It is related to the way of constructing your own identity in interaction with the identities of others and includes the competences of interacting in a broad context, conceiving and conducting personal plans, and accepting decrees, interests, limits and needs.

Finally, DeSeCo proposes a route that allows the estimation of the standards that guide compulsory basic education, in the form of a test designed for three specific areas of knowledge, scientific, mathematical and linguistic, whose results would contribute elements not only on the levels acquired by the students, but also on the conditioning factors of the socio-cultural and family contexts of the same, as shown, for example, in official statements of the Instituto Colombiano para la Evaluación de la Educación (ICFES, 2008). Such is the now very well-known PISA (Program for International Student Assessment) test.

10.2.4 Competences and Scientific Competences in PISA

PISA is a program developed and run by the OECD's educative division:

> [It] aims at providing a new basis for policy dialogue and for collaboration in defining and operationalizing educational goals in innovative ways that reflect judgements about the skills that are relevant to adult life. It provides inputs for standard-setting and evaluation; insights into the factors that contribute to the development of competences and into how these factors operate in different countries, and it should lead to a better understanding of the causes and consequences of observed skill shortages. By supporting a shift in policy focus from educational inputs to learning outcomes, PISA can assist countries in seeking to bring about improvements in schooling and better preparation for young people as they enter an adult life of rapid change and deepening global interdependence. (OECD 2001, p. 3)

Not being this the place to discuss the lights and shadows of the program, it is enough at this point to recall that PISA has both a large number of supporters, as well as several critics. Further discussions from the later perspective could be found in recent works as those of Carabaña (2015), Hargreaves and Shirley (2012) and Fernández, Fernández, and Galindo (2017), while more sympathetic views are found in authors like Spillane and Coldren (2011) and Schleicher (2009). In addition, while there is still no full agreement at the theoretical level about the precise difference between general and specific competences, some authors even unsure about what exactly the later are (Carabaña, 2011; Eslava, 2014), not about the actual role played by schools in their development (Carabaña, 2015). The PISA test refers to specific competences in reading, mathematics and science, detailed and subdivided into sub-competences. In the concrete case of science, a competence is defined in terms of the ability to use knowledge and scientific processes, not only to understand the natural world, but also to intervene in decision making about the use of science and technology. More specifically, scientific competence is:

> The ability to use scientific knowledge to identify problems, explain scientific phenomena and draw conclusions based on evidence on issues related to science, In addition, includes the understanding of the characteristic features of science, understood as a method of knowledge and human research, the perception of the way in which science and technology shape our material, intellectual and cultural environment and the willingness to engage in matters related to science and ideas about science as a reflective citizen. (OECD, 2006, 17)

The above definition involves the following elements:

- The knowledge that students should acquire about scientific concepts, ecosystems, the properties of matter and the transformation of energy.
- The competences that students have to demonstrate for the use and interpretation of scientific evidence and the preparation of predictions in scientific phenomena.
- The contexts in which students encounter scientific problems and the application of knowledge in making personal decisions and understanding global issues.
- The attitudes and dispositions of students toward science.

Being competent in the science area implies, in addition, understanding and manipulating scientific information, acknowledging the nature of scientific knowledge and the limitations that such knowledge has. Science education should encourage students to believe that it can change society and individuals, so PISA aims to evaluate the ability to become socially involved as a citizen, aware of issues dealing with the relations between science and society.

In the natural sciences, PISA evaluates in detail three sub-competences, selected for their importance for the development of scientific practices and for their connection with cognitive abilities. PISA scientific sub-competences (SC) and indicators (I) are as follows:

SC1: Identify scientific issues or topics:

SC1- Ia) recognizes issues that can be scientifically investigated.
SC1- Ib) identifies key terms for the search for scientific information.
SC1- Ic) recognizes the key features of scientific research.

SC2: Use scientific evidence:

SC2- Ia) interprets scientific evidence and prepares and communicates conclusions.
SC2- Ib) identifies the assumptions, evidence and reasoning behind the conclusions.
SC2- Ic) recognizes the social implications of scientific and technological advances.

SC3: Scientifically explain the phenomena:

SC3- Ia) applies knowledge of science to a given situation.
SC3- Ib) describes or interprets the phenomena scientifically and predicts changes.
SC3- Ic) identifies the appropriate descriptions, explanations and predictions.

Scientific competence is evaluated in what PISA calls the dimensions of knowledge and scientific concepts, processes and situations (contexts) with the student's achievement graded in competence levels as follows:

Level 1: Students have limited scientific knowledge that only applies to few situations and give obvious explanations based on evidence.

Level 2: Students give explanations in contexts they know, draw conclusions from simple experiences and literally interpret the results of an investigation or the resolution of a problem.

Level 3: Students identify scientific problems in various contexts, select knowledge to explain phenomena, apply simple research models, use concepts from different disciplines and develop brief statements using knowledge.

Level 4: Students effectively work situations and problems that involve explicit phenomena required to make inferences, select and integrate explanations from different disciplines, reflect and communicate their decisions.

Level 5: Students identify the scientific components of complex situations, compare, select and evaluate scientific evidence to respond to specific situations, construct explanations based on evidence.

Level 6: Students consistently explain and apply scientific knowledge and knowledge about science, demonstrate reasoning and advanced scientific thinking and use scientific knowledge to solve personal, social situations.

Levels, indicators, competences (key, general and specific), subcompetences, scientific and technological development, all of them have a say in the configuration of Colombia's educational policy and, as we are about to see, are entwined across the documents that express and define it.

10.2.5 Studies About Scientific Competences in Colombian Scientific Education Policies

As it will be detailed later, the incorporation of a competence-based framework, and in particular one that includes scientific competences as one of the basic competences in the current Colombian educational policies, has been a process that, far from concluded and consolidated, has been dynamic and controversial. However, once the decisions were made, as mentioned before principally because of the commitment to international treaties and led by the idea of becoming a scientifically skilled nation, the country has never back down in its effort for consolidating its scientific education policies.

The incorporation just referred can be seen, from a chronological perspective, as progressed; that coincides with a quinquennial partition of the temporal framework covered by the documents we studied. In this view, it can be stated that the first period, from 1990 to 1994, sees the initial introduction of the language of competences in policy documents, as well as the defense of their relevance in the educational context. Such is the spirit that holds the 1996–2005 *Plan Decenal de Educación* (Decennial Education Plan) (Ministerio de Educación Nacional, MEN, 1996), in which competences are listed along with abilities, skills and values as educational goals, and the *Lineamientos Curriculares de Ciencias Naturales y Educación Ambiental* (Natural Sciences and Environmental Education Curricular Guidelines) (Ministerio de Educación Nacional, 1998), where the knowledge of the basic scientific concepts and processes is said to help the development of knowledge and actions processes, as well as the competences that characterize the scientific endeavor.

The second interval, from 1995 to 2000, can be defined as the implementation period, when the horizon defined in the first policy documents is to be transformed into concrete actions that allow (would allow), to define the reach and scope of a scientific education based on the notion of scientific competences. The introduction of the *Estándares Básicos de Competencias en Ciencias Naturales y Ciencias Sociales* (Natural Sciences and Social Sciences Basic Competence Standards) (Ministerio de Educación Nacional, 2004), belonging to the first series of basic competence standards defined for all basic subject areas (mathematics, languages and sciences), is considered a corner stone for is quickly became the national referral for an education based in competences, and defined the language, contents and evaluation methods used for both the public and private education sectors.

Finally, the 2005–2010 period, expected to be the time of the consolidation of the competence model, was instead the period where the first systematic evaluations of the model came into place, whose results ended up creating the need for adapting the evaluation model used by the, at the time, only public evaluation agency, the ICFES, Instituto Colombiano para el Fomento de la Educación Superior (Colombian Institute for the Promotion of Higher Education) and for reviewing and intervening on a system that showed no major progress toward the consolidation of a scientifically educated nation.

These same temporal intervals are useful to understand the reactions that came to place from the academy and the teachers' organizations and syndicates, while the scientific competences, actually, the whole competence-based model, were being implemented. Then the first period of implementation was promptly followed by studies that tried to establish the links between the "new" competence frame and some better-known models (psychological, behavioral) in the context of the evaluation systems that were needed for the competence-based, as opposed to the content-based, model (Bogoya, 2000; Gómez, 2004; Torrado, 1998, 2000), and some that tried to measure its impact, or possible impact, at the school level (Cárdenas & Sarmiento, 2000).

During the second period, studies and interpretations of the scientific competences focused on their relation with other competences (Bustamante, 2001; Hernández, 2005), their impact for teachers and educators (Arteta, Chona, Fonseca, Martínez, & Ibáñez, 2002; Gómez, 2004), and it witnessed the birth of enthusiastic interpretative approaches that offered to teach how to take advantage of the new tools to all schools interested in it (Tobón, 2004). Finally, during the convulsed third period of its implementation, the exploration of a variety of classrooms and curricular experiences (Álvarez, 2009; Claret, 2010; Hernández, 2005; Maldonado, 2006; Villada, 2007) served for both friends and foes to defend their viewpoints, claim for, either a return to the old content-based world or jump eyes-closed to the unstoppable future.

10.3 Methodology

The research proceeded using a historical-documentary (HD) methodology for the analysis of the policy documents selected for this study, complemented with a survey that allowed us to obtain an image of the vision Colombian teachers have about the definition, implementation and evaluation of scientific competences. Given that the investigation was focused on a past period of time, the HD methodology offered the obvious advantage of helping make sense of a large number of primary and secondary sources, while at the same time gave us the opportunity to contrast bibliographical material distant in time by the use of common categories. Thus, we developed a view that covers a 15-year period of dramatic changes in Colombian education policy, being able to trace its evolution, and one that allowed us to "connect the dots" that would, at the end, configure an image of the presence, coherence and reach of the official discourse about scientific competences.

On the other hand, the survey developed to communicate with teachers from all the country proved to be a strong instrument for fast and systematic data collection. As usual with this kind of instruments, there is a gap between the expected number of responses and the actual number of them, but these final figures where solid enough to allow a comprehensive view of the way teachers have felt and made sense of a discourse that is sometimes found obscure or incoherent.

The documental analysis advanced in three phases: (a) pre-analysis, (b) review and analysis of documents and (c) data processing and preparation of the final report. The survey started with the drafting and testing of instruments, followed by the invitation to the teachers to participate, the reception of the answers for the final version of the instrument, and their systematization and analysis. Below we offer some details of each of the phases undergone.

Pre-analysis: After establishing a general vision of the problem, a scheme was drawn up to determine the ideas that were intended to be developed. This scheme allowed us to establish limits for data collection and interpretation. During the pre-analysis phase, the documental universe and the units of analysis were defined and tested. The collection and selection of the basic bibliography was accompanied by the elaboration of files where the data of each one of the documents was recorded independently in such a way that they could be identified according to the established categories.

The period of time from where the documents were selected correspond to a very specific moment of Colombia's educational policies. Right after the promulgation of the 1994 General Education Law, a direct heiress of the country's constitutional reform of 1991, a series of policy documents, were introduced in order to contribute a legal ground for the expectations and commitments raised and defined by the new constitution. Consequently, the 1995–2000 interval witnessed the implementation of a large set of documents aimed at introducing the new language of the competences, as well as the definition of specific competences for the basic areas of the national curriculum.

In the other side of the temporal spectrum, 2006 was the year in which Colombia participated for the first time in PISA's international evaluation, focused on the measurement of scientific competences—entering in an era of permanent and systematic measurements and evaluations, both of the students' academic performance and of the nation's educational system—in order to better their results, of the students and of the general system, as well as the country's options for being accepted into the club of the OECD countries/economies. As a result, between 2008 and 2010, the ICFES conducted a review exercise of the census-based tests carried out in the country that resulted on a major change of the instruments, evaluation models, information to be collected, and, above all, conceptual horizon, moving from a traditional memory-based one to a competence-based one.

At the end, the documental universe was made up of more than a hundred official documents produced by the Ministry of Education and its affiliated entities, between 1995 and 2010. The nature of the documents included laws, norms, decrees, resolutions, regulations, and conceptual and policy guidelines. Only officially published documents were reviewed, and although we had access to a number of them, drafts, or not authorized versions of documents, were not taken into account.

Review and analysis of documents: The review of the documents was carried out in two moments: categorization and coding. In the first of them, the units of meaning, assigned to each unit of registration, were classified, and a system of analysis categories was determined. This first step included an initial reading of the selected documents in order to determine if the selected sample is suitable to be subjected to analytical procedures. During the second moment, the coding built the explicit categories of classification of documents, and, after a process of decomposition, aggregation and enumeration, the representation by clusters of the content of each document was constructed.

In the analysis phase, the information was encoded, decomposed and enumerated, registering it appropriately in the log sheets designed for it. The sheets contain the units of analysis, in which elements that characterize the material under analysis, as, for example, name, publication date, authors, location, page and section, were included. A record sheet was designed for each document in order to have independent data from each of the analysis units as well as a way to determine if each document directly or indirectly refers to the concept of scientific competence, either by semantic or conceptual affinity. The results were quantified according to their usefulness for the identification of the concept of scientific competence, allowing us to build reference groups with similar uses of the semantic units, where the concepts and documents in which the units were present. This reference group was called a cluster.

For data processing, statistical operations and synthesis and selection of results were carried out. A general analysis of the qualitative dimensions of the documents was also carried out, where the documents were organized in relation to the frequency of appearance of the categories, expressing them in percentages. These percentages were ordered from highest to lowest to establish contrasts and regularities; similarities and differences between sets of documents.

Within the strategies of content analysis of this work, a survey has been introduced with the purpose of carrying out an exploration of teachers' conceptions related to scientific competences. For this, a questionnaire about the characterization of scientific competences, was designed and shared using the Google Docs survey tool. The survey allowed us to understand, to some extent, the scope that educational policy documents based on competences have had within the teaching community, as it inquires teachers' knowledge about the concept of scientific competence present in official documents and how this knowledge is reflected in the classroom practices. In addition, it offers tools to establish the internal consistency of the strategies proposed in official documents over time, since it is the teachers who use these frames of reference. Finally, since it inquires explicitly about the knowledge of legal references, it is a tool that allows gathering information on the effectiveness of the transmission of relevant information by the entities that generate official reference documents.

10.4 Results

The outcomes of the study are now presented and organized according to three major categories. First, there is the general statistics category, where we report the more global findings, regarding the notions, actions and strategies concerning how a scientific competence-based education is to be defined an implemented. The second category, cluster analysis, deals with the internal coherence of policy documents, as with consistency issues among them. The final category, survey, reports its results, where the voices of Colombian educators are the main character.

10.4.1 General Statistics

- More than a half of the documents reviewed were located in the categories of guides and documents, which are characterized by presenting pedagogical and didactic orientations. However, only in the most recent of them, there are elements that allow for a clear characterization of the competences. Most of them, in particular those prior to 2005, do not include an explicit definition of what a competence is, even though, given their nature, they are designed to offer guidelines of implementation of scientific competences. Only an implicit notion of the competences can be found in these documents, addressing, instead, topics such as objectives or training purposes of natural science curriculums, without any further development of the notion of competence or that of competence-based education.
- In relation to the concept of scientific competence, in a very few of the documents were found elements to explain how scientific competences develop as part of concrete instructional practices or how evaluations of student's learnings may determine the level of achievement of scientific competences. Additionally, the elements that define and categorize the scientific competences are explicitly found mainly in documents that accompany the presentation and interpretation of census test results, that is, tests applied to all the students in the system, which shows that the most frequent conceptualizations of competences are those associated to their external evaluation.
- Regarding the presence of activities that encourage the development of competences, only a small percentage of the documents present any concrete example of it, very similar to the number of documents where there is a mention of the role of teachers as helpers in the process of development or improvement of competences in their students.
- As a result of the general statistical analysis, it can be stated that there seems to be a lack of coherence between a conceptual competence-based approach, that guides the formulation of a national framework for a robust scientific education, and the pedagogical practices that would enforce it, both at the theoretical-interpretative and the practical-pedagogical levels.

10.4.2 Clusters Analysis: Internal Coherence and Consistency

- From the analysis of the clusters we determined that the concept of scientific competence, as is also the case for all the basic competences, includes a comprehensive performance in the execution of actions, the assumption that the skills acquired through this execution are linked to productivity and work, and the existence of an accompanying conceptual knowledge. Formulated in such a general way, the concept of competence presented in the documents does not seems to be particularly different from the common definition that was given to intelligence in the educational models based on psychology, as the models developed during the first half of the twentieth century. As a consequence, the policy documents tend to equate scientific competences with a combination of formal and practical intelligence, making it difficult to understand not just the different perspectives from which each model, intelligence and competences, comes from, but also the conceptual differences underlying their languages. It does not come as a surprise then, as it was later the case in the responses to the survey, that some teachers consider that the move from a content-based to a competence-based approach to science education is nothing but a change in words.
- The notion of competence that appears in the documents breaks with the idea that education must attend to something more than the transmission of knowledge and states that the most important thing is the development of thinking skills, which become comprehensive learning of a general nature and are expressed in a multiplicity of situations and contexts. Several documents maintain that the competences involve the cognitive, attitudinal and procedural dimensions (Anderson & Krathwohl, 2001). This is clearly reflected in the structure of the basic standards of natural and social sciences, where the three dimensions are put into play in an independent and integrated way.
- Talking about scientific competences only makes sense from a perspective of a science that is lived, that is constructed and that changes with the rhythm of human intentions; it is not possible to isolate the scientific knowledge of life, its applications, its implications and its relations with other areas of knowledge. Science does not begin with facts but with questions, and questions depending on the theoretical framework from which they are formulated. In this sense, between

the curricular guidelines and the basic competency standards there is a common thread that assumes that the purposes of science education, in secondary education, should focus on a didactic aimed at solving problems, extracted from daily life and that empowers the competences more than in factual knowledge. In this sense, the policy documents are fully consistent with the recommendations resulting from contemporary research in science education.

- Although there is a consensus on the importance of improving quality, in terms of knowledge rather than competences, there is no evidence of the way in which all the transformations required to promote this change (solving problems such as poverty, violence, destruction of the environment) are powered by socialization processes in schools. Making it more a matter of personal performance than of contextualized social interaction. Therefore, while following the discourses and ideals of international agencies, there is nothing in the Colombian policy that would guide interested stakeholders to implement them in the particular conditions and attend to the specific needs of the country.
- In most of the documents evaluated, that were written after the turn of the century, there is a recurrent idea of establishing educational goals that point to the formation of integral individuals who participate actively and constructively in their world. The achievement of these goals requires a structured and goal-oriented organization approach to the scientific competences and contents to develop and teach, in order for them to be integrated into a broader overarching framework that includes explicit addressing of conceptual knowledge (know-what), procedural knowledge (know-how), attitudinal knowledge (knowing-together) and knowledge to learn (know-to-know). However, explicit characterizations of the scientific competences were only found in the texts of the theoretical framework of the census tests produced by the ICFES, where the ethical, economic, social and environmental dimensions, which must accompany the construction of scholarly scientific knowledge, are left aside. Furthermore, most of the documents with an explicit reference to competences state that working out and evaluating competences requires organizing the curriculum and assessments by levels of increasing complexity; these necessarily implies changes in the aims of education, which, as presented, does not show articulation between the different educational levels.

- In the same vein, it was found that the documents produced between 1994 and 2004 focused their attention on the transmission of knowledge. The competence approach that appears in latter documents is related to job training, in an attempt to build bridges between the productive and the educative sectors, something that gives competences a work-force taint, where finer nuances must be incorporated in order to guarantee that competences become linked with an integral education, integrating job skills but also social attitudes and values as empathy, responsibility and caring.
- Several approaches to the conceptualization of scientific competences are evident, too many to allow a clear path to a common understanding. There are behavioral approaches, with competences seen as key behaviors of people who are brought into play in challenging circumstances; functionalist approaches, where competences are presented as a set of attributes that people should have to comply for labor and professional purposes; constructivist approaches, in which there seems to be little difference between skills and competences, at least beyond the easy fact that the later are nothing more than the former when applied to concrete contexts and situations. Finally, there is what can be named as a competitivity approach to competences, a view that sees competences as the trading currency in, present or future, labor contexts, with the additional clause that only those demonstrating eagerness for success, even if it is at the expense of others, can triumph. It would be necessary for a solid and coherently designed educational policy to take a stand and settle the interpretative issues, in order to create an environment where all stakeholders find common ground to start the construction of a new kind of citizenship, scientifically knowledgeable and accountable.
- The categorization of scientific competences defined in educational policy documents is highly influenced by those defined in the PISA program, especially in relation to the description and explanation of phenomena and the identification of scientific issues. They differ mainly in the reflection that PISA proposes about the social implications of scientific and technological advances as an area of evaluation, the use of scientific evidence and the elaboration of conclusions and aspects related to scientific activity, implicitly stated in the guidelines' curricula and basic standards of competences. It is also found that there is a gap in relation to other frames of reference (international guidelines for census tests). These frameworks affirm that for the

identification and characterization of scientific competences it is necessary to establish as a priority the issues that may be identifiable from the science, search for information and approach the problem from the experimental perspective. The documents reviewed do not address the identifiable scientific issues and do not make any emphasis on scientific communication, forgetting that the ultimate goal of a competence-based scientific education is to allow students to advance in the processes of consolidation of scientific literacy and the use of such knowledge under the contextual conditions of each individual.

- Due to the absence of a link between the competences approach with pedagogical theories, there are problems in the approaches that guide the formation of teachers. The landing of the approach requires profound changes in the teaching practice and the structuring of the curriculum. For them the training processes (which are usually slow, inadequate and ineffective) must be given in terms of the educational level in which they perform. The training of teachers should be given in three lines: the disciplinary, pedagogical and professional-labor; related in such a way that would allow the understanding of education as an engine that powers multidimensional relations between human individual, communities and environments.

10.4.3 Survey: The Vision of the Colombian Teachers and Experts

- The instrument and the methodology allowed communication with a high number of teachers, from which only a small percentage of in-service teachers and experts responded. However, beside the responses to the survey, a high number of emails were received. Some of these messages manifest interest in the subject, but most of them express dissatisfaction and distrust in the face of carrying out an investigation of this type, supported by arguments of a political and economic nature.
- The mechanism for collecting information, through an electronic application, is quick and useful to know the perceptions of teachers in relation to the subject of competences and can be consolidated as a valid tool to request and access information from different institutional management areas. Having had the participation of teachers

from all the country's departments (states) and from nearly cities, makes it possible to demonstrate that the way in which science education is taught, the pedagogical practices that accompany it and the selection of resources, are determined by both institutional and contextual factors.

- Few institutions have their own plans and programs for the area of natural sciences. In general, the contents and organization of the areas are determined by the knowledge and skills of the teachers, who in some cases do not have specific training on the subject. The pedagogical practices tend to be oriented toward the labor competences demanded by the context or region in which the population is located; this is how in some cases, the contents of the area are defined in conjunction with other municipal entities. In other cases, the teachers determine the classroom plan according to the economic conditions and access to different resources in the region, that is, what is done in the science class depends on the conditions of the students and teachers.
- The current number of educators in charge of teaching science is low in relation to the number of seats and students that are currently in educational institutions. In addition, some of the teachers in charge of scientific disciplines do not have specific training in the area, even less any knowledge in the policy documents that guide their practice.
- In a high percentage, the respondents state that they have found the concepts of competence and scientific competence in the documents. However, when answering questions about the opinion that the concepts deserve, most of the teachers declare not to have enough basis, or knowledge of the legislation, to offer a sound answer.
- Only a very small percentage of respondents stated that the selection of the contents to be taught in their classes is actually done taking as a reference the development of scientific competences. Most of the teachers declared that this selection is made based on the basic competency standards, the curricular guidelines and the textbooks, but they hardly go beyond repeating what they find there. It is then necessary to reflect on the understanding, interpretation and appropriation that is made of these and other documents that guide the definition of contents under a logical sequence of learning processes in order to achieve significant pedagogical practices in the teaching of science.

- Finally, the main pedagogical strategies used by teachers in charge of teaching science indicate an emphasis on procedural and production skills, rather than in any sort of integrated competence-based approach.

10.5 Conclusions

The analysis allows us to conclude that, from the year 2000, the concept of scientific competence starts to appear recurrently on documents onward, making it evident that the introduction of the competence framework for scientific education took a time to become embedded in the national policies. Post-2005, documents do not develop any further the competence framework, but rely on those of the previous period, making the 2000–2005 policy documents the basic guides for all the stakeholders nationwide.

The impact of the competence-based scientific education is perceived in educational policy framework through their positioning as the product of educational processes that guarantee the achievement of scientific educational goals. Then, it became a fundamental task to guarantee that a common language was defined for, and used by, institutions, educators, students and families, in order to facilitate its implementation all over the national education system. Achieving such a goal is considered to be a fundamental step toward the bettering of the system as a whole, as to guarantee that all students have a fair opportunity to develop those competences, not only to respond to the demands of a globalized economy but mainly to offer equal conditions to ethnic and cultural diversity.

Knowing that the concept of scientific competence is a dynamic one and that its articulation with the general competence-based framework is a must, in order to consolidate a solid national educational policy, an overarching model for the development of competences in Colombian educational context must be advanced. Such a model, as it needs to respond to the demands of different stakeholders, would have to be construed after an open dialogue between teachers, teacher associations, families, disciplinary and pedagogical experts, government representatives. Only a consensual model would guarantee that all parties commit to its implementation and protection from the external forces that could affect it.

The elaboration of programs based on the development of scientific competences requires, among others, the construction of a shared conception of the notion of competence, and of competence-based learning and teaching. The definition of the conceptual and attitudinal resources that will mobilize the competences, that is, knowledge and attitudes, must be included in science curricula all across the educational cycles, from preschool to high school, guaranteeing the didactic approaches are appropriate to the orientations and aims of each level. Particular attention should be given to the selection of evaluation formats and activities that allow for a clear understanding of the students advance in the development of their competences, and to the implementation of a mechanism that warrants that teachers and educators gain ownership of the scientific competences framework and make use of it in a permanent and consistent way.

The postulates presented in the documents emphasize the performances as means of representing the state of the competences in the individuals. These performances require processes of internal transformation, differentiated according to the type of competence, its complexity and the learning environments that would allow their development. Therefore, it is essential to understand that only from a deep and up-to-date knowledge of the core elements of scientific competency-based training can pedagogical practices be designed to promote and stimulate them.

The competency approach is proposed as an alternative for curricular design and for the development of teaching–learning processes; from a constructivist perspective, the development of competences is considered as know-how in practice, motivated by meaningful learning and transferred to real-life situations that involve solving problems. In this sense in Colombia there is much to be defined and investigated to clarify and differentiate between alternative interpretations of the theoretical framework devised for the development of competences. At this point it is important to understand that scientific competences to be acquired by the students are not an automatic consequence of the design of effective learning activities and changes in the execution of traditional tasks, but of the establishment of an effective link between goals, performances, attitudes, problem-solving opportunities and application contexts.

References

Álvarez, M. (2009). *El desarrollo de competencias científicas a través de una experiencia de introducción de la historia de las ciencias y las técnicas en el aula.* Enseñanza de las Ciencias, 2186–2192.

Anderson, L. W., & Krathwohl, D. R. (Eds.). (2001). *A Taxonomy for Learning, Teaching, and Assessing: A Revision of Bloom's Taxonomy of Educational Objectives.* New York: Addison Wesley Longman.

Arteta, J., Chona, G., Fonseca, G., Martínez, S, & Ibáñez, S. (2002). Las competencias científicas y el pensamiento de los profesores de Ciencias Naturales. En B. Barbosa (Ed.), *El Oficio de Investigar. Educación y Pedagogía Frente a Nuevos Retos.* Bogotá: CIUP-Universidad Pedagógica Nacional.

Bogoya, D. (2000). Una prueba de evaluación de competencias académicas como proyecto. In D. Bogoya et al. (Eds.), *Competencias y proyecto pedagógico.* Bogotá: Universidad Nacional de Colombia.

Bustamante, G. (2001). La competencia lingüística, ideológica y comunicativa, a propósito de la evaluación en Colombia. En SOCOLPE, *El concepto de competencia: Una mirada interdisciplinar.* Bogotá: SOCOLPE-Alejandría.

Carabaña, J. (2011). Competencias y universidad, o un desajuste por mutua ignorancia. *Bordón, 63*(1), 15–31.

Carabaña, J. (2015). *La inutilidad de PISA para las escuelas.* Madrid: Catarata.

Cárdenas, F., & Sarmiento, F. (2000). *Competencias y proyecto pedagógico. Desarrollo y evaluación competencias en Ciencias.* Bogotá: Universidad Nacional.

Care, E., & Luo, R. (2016). *Assessment of Transversal Competencies: Policy and Practice in the Asia-Pacific Region.* Paris: UNESCO.

Claret, A. (2010). *Educación y formación de competencias en ciencias naturales.* Bogotá: ASCOFADE.

Comisión de las Comunidades Europeas. (2005). *Propuesta de recomendación del Parlamento Europeo y del Consejo sobre las competencias clave para el aprendizaje permanente. 2005/0221(COD).* Bruselas: Comisión de las Comunidades Europeas.

Consejo de Europa. (2000). *Conclusiones de la Presidencia, 23 y 24 de marzo de 2000.* Lisboa: Consejo de Europa.

Consejo de Europa. (2002). *Conclusiones de la Presidencia, 15 y 16 de marzo de 2002.* Barcelona: Consejo de Europa.

Delors, J., et al. (1996). *Learning: The Treasure Within. Report to UNESCO of the International Commission on Education for the Twenty-first Century.* Paris: UNESCO.

Eslava, E. (2014). Conceptos, pertinencia y acceso: una mirada filosófica a la educación en ciencias naturales. *Revista Colombiana de Filosofía de la Ciencia., 14*(28), 67–82.

Eslava, E. (2015). Educación en América Latina: retos y oportunidades para la filosofía de la región. *Universitas Philosophica, 32*(65), 223–244.

European Parliament. (2006). *Recommendation of the European Parliament and the Council. (2006/962/CE).* Brussels: European Parliament.

Faure, E., et al. (1973). *Learning to Be. The World of Education Today and Tomorrow.* Paris: UNESCO.

Fernández, C., Fernández, O., & Galindo, E. (2017). *Escuela o barbarie. Entre el neoliberalismo salvaje y el delirio de la izquierda.* Madrid: Akal.

Figazzolo, L. (2009). *Testing, Ranking, Reforming: Impact of PISA 2006 on the Education Policy Debate.* Brussels: Educational International.

Giroux, H. A., & Searls, S. (2004). *Take Back Higher Education: Race, Youth, and the Crisis of Democracy in the Post-civil Rights Era*. New York: Palgrave Macmillan.

Gómez, J. (2004). Lineamientos pedagógicos para una educación por competencias. En SOCOLPE, *El concepto de competencias II: una mirada interdisciplinar*. Bogotá: SOCOLPE-Alejandría.

Halinen, I. (2017). *The Conceptualization of Competences Related to Sustainable Development and Sustainable Lifestyles*. Paris: UNESCO.

Hargreaves, A., & Shirley, D. (2012). *The Global Fourth Way: The Quest for Educational Excellence*. Thousand Oaks: Corwin Press.

Hernández, C. (2005). *¿Qué son las competencias científicas? Foro Educativo Nacional. Colombia*. Bogotá: Magisterio.

Hirmas, I. (2009). *Educación y Diversidad Cultural: Lecciones desde la práctica innovadora en América Latina. Red Regional de Innovaciones Educativas para América*. Santiago: OREALC/UNESCO.

ICFES. (2008). *Colombia en Pisa*. Bogotá: ICFES.

Maldonado, M. (2006). *Las competencias, una opción de vida: metodologías para el diseño curricular*. Bogotá: ECOE.

Ministerio de Educación Nacional. (1996). *Plan decenal de educación 1996–2005*. Retrieved from www.mineducacion.gov.co/1621/articles-85242_archivo_pdf

Ministerio de Educación Nacional, 1998. Lineamientos Curriculares Ciencias Naturales y Educación Ambiental. Ministerio de Educaión Nacional: Bogotá.

Ministerio de Educación Nacional. (2004). *Estándares Básicos de Competencias en Ciencias Naturales y Ciencias Sociales. Serie Guías, No. 7*. Bogotá: Ministerio de Educación Nacional.

OECD. (2001). *Definition and Selection of Competences. Theoretical and Conceptual Foundations. Background Paper*. Paris: OECD.

OECD. (2002). *Definition and Selection of Competences. Theoretical and Conceptual Foundations. Strategic Paper*. Paris: OECD.

OECD. (2003). *Definition and Selection of Key Competences. Executive Summary*. Paris: OECD.

OECD. (2006). *El programa Pisa de la OCDE. Qué es y para qué sirve*. Paris: OECD.

OECD. (2007). *PISATM 2006 Science Competences for Tomorrow's World*. Paris: OECD.

Roegiers, X. (2016). *Marco conceptual para la evaluación de competencias*. Paris: UNESCO.

Rychen, D. S., & Salganik, L. H. (Eds.). (2003). *Key Competences for a Successful Life and a Well-functioning Society*. Seattle, WA: Hogrefe.

Schleicher, A. (2009). International Assessments of Students Learning Outcomes. In L. Pinkus (Ed.), *Meaningful Measurement: The Role of Assessment in Improving High School Education in the Twenty-first Century*. Washington, DC: Alliance for Education.

Spillane, J., & Coldren, A. (2011). *Diagnosis and Design for School Improvement: Using a Distributed Perspective to Lead and Manage Change*. New York: Teachers College Press.

Stiefel, B. (2008). Competencias Básicas. In *Hacia un nuevo paradigma*. Madrid: Narcea.

Tobón, S. (2004). *Formación basada en competencias*. Bogotá: ECOE.

Torrado, M. (1998). *De la evaluación de aptitudes a la evaluación de competencias: aportes de la psicología a la reconceptualización del examen de estado*. Bogotá: ICFES.

Torrado, M. (2000). *Competencias y proyecto pedagógico. Educar para el desarrollo de las competencias: una propuesta para reflexionar*. Bogotá: Universidad Nacional.

UNESCO. (1966). Recomendación conjunta de la OIT y la UNESCO relativa a la situación del personal docente. En: UNESCO y Organización Internacional del Trabajo OIT. (2008). *Recomendación conjunta de la OIT y la UNESCO relativa a la situación del personal docente y Recomendación de la UNESCO relativa a la condición del personal docente de enseñanza superior*. Paris: UNESCO.

UNESCO. (1997). *Defining Tolerance*. Paris: UNESCO.

UNESCO. (2000). *Science for the Twenty-first Century a New Commitment. Declaration on Science and the Use of Scientific Knowledge. Science Agenda – Framework for Action*. Paris: UNESCO.

UNESCO. (2002). *Declaración de Sao Pablo. Conferencia Interparlamentaria de América Latina y el Caribe*. Paris: UNESCO.

UNESCO. (2004a). *Educación para todos: el imperativo de la Calidad. Informe de seguimiento de la ETP en el mundo*. Paris: UNESCO.

UNESCO. (2004b). *Changing Teaching Practices; Using Curriculum Differentiation to Respond to Students' Diversity*. Paris: UNESCO.

UNESCO. (2004c). *Declaración de Buenos Aires. II Reunión de ministros del Comité Intergubernamental del Proyecto Regional de Educación para América Latina y el Caribe PRELAC*. Paris: UNESCO.

UNESCO. (2005a). *Hacia las sociedades del conocimiento. Informe mundial de la UNESCO*. Paris: UNESCO.

UNESCO. (2005b). *Recomendaciones del Parlamento Europeo y del consejo sobre las competencias clave para el aprendizaje permanente. Bruselas*. Paris: UNESCO.

UNESCO. (2007a). *Educación de Calidad para Todos. Un asunto de derechos humanos. Políticas educativas dentro del marco de la II Reunión Intergubernamental del Proyecto Regional en Educación para América Latina y el Caribe (EFA/PRELAC)*. Santiago: UNESCO.

UNESCO. (2007b). The State of Education in Latin America and the Caribbean: Guaranteeing Quality Education for All. In *A Regional Report, Reviewing and Assessing the Progress Toward Education for All Within the Framework of the Regional Education Project (EFA/PRELAC)*. Santiago: UNESCO.

UNESCO. (2011a). *UNESCO ICT Competency framework for teachers.* Paris: UNESCO.

UNESCO. (2011b). *Déclaration de Busan. Premier Forum mondial sur les humanités.* Paris: UNESCO.

UNESCO. (2014). *Declaración de Lima. Educación para Todos (EPT) en América Latina y el Caribe: Balance y Desafíos post-2015.* Paris: UNESCO.

UNESCO. (2015). *Mirando hacia 2030: fortalecer el desarrollo de competencias laborales, medio de un mejor futuro para todos.* Santiago: OREALC/UNESCO.

Villada, D. (2007). *Competencias.* Manizales: Sintagma.

Winterton, J., Delamare-Le Deist, F., & Stringfellow, D. (2006). *Typology of Knowledge, Skills and Competences: Clarification of the Concept and Prototype.* Luxembourg: Office for Official Publications of the European Communities.

CHAPTER 11

Perception in Science, Technology and Innovation: A Comparative Review for the Years 2012 and 2015 from Colombian Surveys

Sylvia Patricia Fletscher Moreno and Clara Inés Pardo Martínez

11.1 Introduction

Surveys of the public perception of science and technology (SPPS&T) emerged in the United States in 1957. The use of these instruments in Latin America dates to 1987, when they were first implemented in Brazil; in Colombia, they were applied for the first time in 1994. In addition, over the last decade, various countries in the region developed this type of survey based on the support and articulation of the Network for

S. P. Fletscher Moreno (✉)
Colombian Observatory of Science and Technology (OCyT), Bogotá, Colombia
e-mail: sfletscher@ocyt.org.co

C. I. Pardo Martínez
School of Management, Universidad del Rosario, Bogotá, Colombia
e-mail: clara.pardo@urosario.edu.co

C. I. Pardo Martínez et al. (eds.), *Analysis of Science, Technology, and Innovation in Emerging Economies*,
https://doi.org/10.1007/978-3-030-13578-2_11

Science and Technology Indicators—Ibero-American and Inter-American (RICYT). In 2015, RICYT presented the Antigua Manual as a technical proposal with a methodology and practical recommendations for applying national surveys on the public perception of S&T developed by the national agencies for science and technology of Spain and Latin America, with the aim of obtaining comparable results among countries and contributing to the discussion on this topic.

Perception surveys question the interests, attitudes and values about S&T, the social appropriation of science and technology (SAS&T), and participation and public policies in S&T to understand the representations and social developments that count more and predominate among public opinion on these issues. These types of surveys allow the perceptions about science and technology to be considered at a moment in time and over a specific population.

Perceptual studies are an indicator of social realities, with some limitations, such as they cannot sufficiently account for forms of knowledge appropriation, not only because of the type of questions (closed questions) but also because they are circumscribed in a context (e.g., country, place, group of people). For perception surveys in science and technology, as noted by Daza-Caicedo et al. (2009, p. 24), "polls tell us more about the public and publicized aspects of science and technology than about the uses individual of these"; nevertheless, these tools offer an approach to society's conception of these subjects.

Although perception surveys have limitations, perception indicators allow comparative studies to develop over time in the same context or between countries that apply these same instruments, thus allowing policymakers to address needs on attitudes and valuation, appropriation and citizen participation in science and technology.

In Colombia, four surveys on public perception of science and technology (SPPS&T) were conducted, in 1994, 2004, 2012 and 2015. Moreover, the national policy of the SAS&T was launched in 2005, and its fifth section indicates the importance of regularly conducting national surveys on the public perception of S&T, which allowed the application of the last two surveys (2012 and 2015). These surveys used a similar methodology and questionnaire, with the aim of achieving comparability and determining the trends in science and technology perceptions as well as the validity of the instrument.

The main objective of this chapter is to analyze the results of the last two national SPPS&Ts conducted in Colombia in 2012 and 2015, taking into account the fact that surveys on public perception are important because they show the social conceptions that are the most burdensome, the most common, and the most available among the population. These studies allow questions such as the following to be analyzed from different perspectives: what are the main sources from which techno-scientific developments are built; whether the population knows and uses local science and technology; the expectations of scientific communities; and the levels of ownership of science and technology in everyday life. The findings of this study are important for policymakers and scientific communities to strengthen the dialogue with different social stakeholders.

This chapter is divided into four sections. After the introduction, the research method of the study is elucidated. The following section presents the main results and implications obtained from our analysis. Finally, a conclusion about the main findings of this chapter is presented.

11.2 Research Method

The objective of the Colombian SPPS&T (2012 and 2015) is to identify the population's opinions and attitudes on S&T and to generate knowledge on the SAS&T processes in the country. The scope of both studies was the population of 32 Colombian capital cities more than or equal to 16 years of age. The questionnaires included the same questions, except for five new questions that were included in 2015, for a total of 79 questions. Table 11.1 presents the structure of the survey and the number and features of the questions in every module following the Antigua Manual. The survey conducted in 2015 and was based on a methodological design, whose distribution by socio-economic stratum was based on the number of electric customers (according to SIU 2015).

The research design was non-experimental cut. For both years (20012 and 2015), the survey was chosen to collect information because it allowed a quantitative description of the trends, attitudes and/or perceptions of the population to be studied. The sampling design in both studies was probabilistic and multistage stratified, which allowed the results to be inferred and sampling errors to be estimated. From the survey results, 12 indexes were built to synthesize results in similar topics according to socio-demographic variables using two techniques: Multiple Correspondence Analysis (MCA) and K-means.

Table 11.1 Structure of survey on public perception of S&T

Module	*Number of questions*	*Features of main questions*
Socio-demographic characteristics	10	Age, sex, educational level, home city, determination by physical or cultural traits, religiousness, occupation, among others.
Interests and information on S&T	18	Where and how they are informed? How people have learned about S&T as a product of the amount of information they have access to, their interest in consuming it etc.
Attitudes and valuation	18	Cultural indicators that tell us what representations and values around S&T prevail over others and how they change over time.
Social appropriation of S&T	9	The use of information in S&T in everyday life, the actions that citizens take in specific situations related to S&T, the consumption and use of techno-scientific knowledge, among others.
Public participation	4	The willingness to participate in decision making on issues that citizens deem relevant in terms of S&T according to mechanism of citizen participation.
Public policies	20	The levels of science's institutionalization in the country and the extent to which people recognize the institutional spaces of production and management of S&T in the public domain.

Source: OCyT

The indexes proposed are the following: perception index of science, technology and innovation; media index related to science; media index related to technology; space use index for science communication; index of appropriation of science and technology on everyday life; index of information related to medicine and health; index of information related to environment and ecology; index of interests in science, technology, environment and health; participation index; index of interest and benefits of science, technology and innovation; perception index of science education; and perception index of scientific vocation (Daza-Caicedo et al., 2014).

The application of the information collection instrument used different steps for both surveys: training of the personnel who carried out the surveys (delivery of information gathering manual, delivery of data and forms); field work, in which the respondents were selected, taking into account a filter that allowed their adequate choice; the polling firms that validated the information in terms of consistency; and digitization and processing, during which the polling companies performed information digitization for processing in a database.

Using the results of the III and IV national SPPS&Ts, this study conducted a comparative analysis to determine the trends and dynamics on the public perception of S&T in Colombia and to determine the similarities and differences between the studies as well as validity of the instruments.

11.3 Results and Implications

SPPS&Ts determine the interest of population in S&T and identify cultural habits that give an account of SAS&T in everyday life, citizen participation and the perception of the usefulness of S&T and innovation in the country. The results of both surveys show similar trends, which indicates that Colombian citizens have preferences for the use of space for the communication of science, such as museums, botanical gardens and libraries, which allow the processing of SAS&T to increase in daily life. Other results indicate a higher visibility of activities such as reading, proactive searching for technical and scientific data on diverse products and services used in everyday life (e.g., home appliances, bills, drugs), shifts in consumption and lifestyle to counter environmental problems and participation in collective actions that seek to generate changes in the society-nature relationship.

Concerning the question of whether the respondents consider that science is performed in Colombia (see Fig. 11.1), there was a decrease of more than 22% between the years 2012 and 2015 in the number of respondents who affirmed that science is performed in the country. The number of respondents who said that in Colombia is not science increased by more than 13% in the years in question. In addition, the number of Colombians

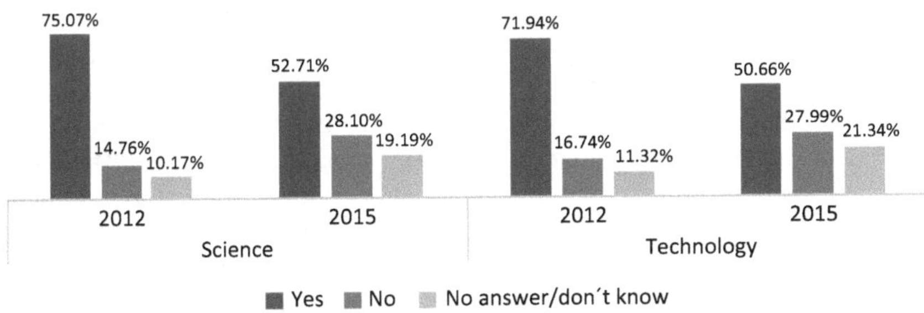

Fig. 11.1 Do you consider that science or technology is made in Colombia? Source: SPPS&T, 2012 and 2015 OCyT

who said they did not know increased from 10.17% to 19.19% between 2012 and 2015. For the respondents who consider that technology is being conducted in Colombia, a behavior similar to that reflected for science was observed relative to those who answered affirmatively (a reduction of more than 21% was registered between 2012 and 2015), who gave a negative answer to this question (an increase of more than 11% between 2012 and 2015) and those who did not know about the subject (increase of 10% between 2012 and 2015). We see how the two indicators report a similar behavior: according to the data, fewer people consider that science and technology are being developed in Colombia, which likely indicates that in the last survey, the attention and interest of population were related to other topics or interests that did not focus on S&T.

11.3.1 Socio-demographic Characteristics

The survey was conducted in the same cities in both years. The sex distribution in the two cities coincided, so the number of women was higher in both years. In 2015, most of the people surveyed were between 26 and 30 years (14.22%), while in 2012, the respondents were between 16 and 20 years (14%).

Regarding the educational level, the maximum level of the respondents' training was basic secondary in both years: 45.18% in 2012 and 45.29% in 2015. This was followed by respondents with a level of basic primary studies, with 22.51% in 2012 and 16.45% in 2015, and by respondents with a university education corresponding to an undergraduate degree, with 14.96% in 2012 and 13.99% in 2015.

11.3.2 Interest and Information S&T

The media's effect on people's opinion is fundamental for explaining or finding actions that solve deficiencies for "correct understanding". In the two years, respondents said that television was the most chosen means of communication as an information system (national television was the most watched by Colombians, followed by subscription television). Unlike 2012, in 2015, respondents said that the second most used medium by Colombians was not the radio but the internet (social networks were the pages most consulted for their content, followed by search engines and directories). In terms of the means by which respondents reported on

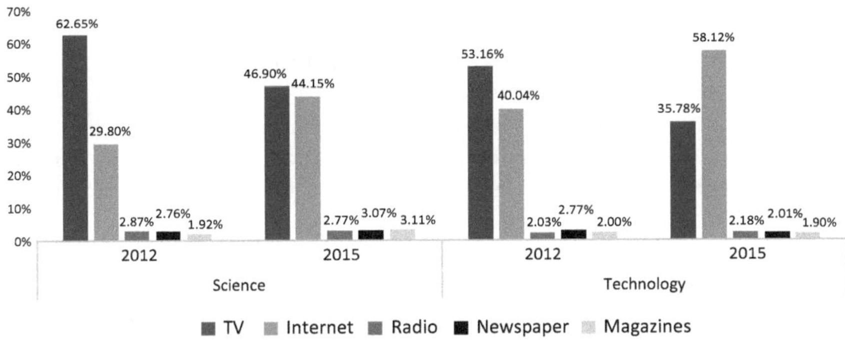

Fig. 11.2 Media by which the respondent is informed about science and technology. Source: SPPS&T, 2012 and 2015 OCyT

technology, in the III SPPS&T, television was the most widely used means of obtaining information on a subject (53.16%), followed by the internet (40.04%). Meanwhile, in the IV SPPS&T, a significant change was evident: in 2015, the internet was the most used medium for information on technology (see Fig. 11.2). These results indicate that the internet was the preferred information system due to its versatile use and presentation of information in real time.

For the two years in which the survey was conducted, people felt very well informed about chronic diseases such as hypertension and diabetes (18.12% in 2012 and 16.63% in 2015), followed by issues related to nutrition and food, with 14.24% in 2012 and 14.83%. For subjects related to people's mental health, therapies and alternative medicine, a very small percentage of people reported being very knowledgeable about the subject.

On the other hand, in relation to information that the respondent had in terms of environment and ecology, people stated that they felt informed about the efficient use of water in both years: 15.27% in 2012 and 19.08% in 2015. A low percentage of respondents reported being informed about the impact of agro-fuels for both years.

The results obtained in the 2012 and 2015 surveys in terms of interests and means highlighted media science's role, which is seen on television and the internet. Even by 2015, the gap between abstract science (from the Manual) and media science (seen in news, cable television etc.) prevails. Moreover, people every day use the internet more to consult on topics related to health and the environment.

11.3.3 Attitudes and Valuation

The correlation between knowledge and attitudes does not necessarily imply a causal relationship between them, since support for science has led to the search for new knowledge. However, having greater knowledge can mean having a greater awareness of the risks and uncertainties associated with particular scientific issues; therefore, it can lead to potentially less support or confidence in science.

As for people's interest in science and/or technology, between 2012 and 2015, the percentage of people who said they were interested in science and/or technology decreased from 74.27% to 64.26%. The percentage of respondents who said they were not interested in science and/or technology increased from 25.73% in 2012 to 35.74% in 2015. These results could be derived from the respondents' unawareness regarding the role and importance of science and technology for development and welfare because its relationship between labor and daily activities was unclear.

For the respondents' perception of the characteristics that describe a person who does science, for the two survey years (2012 and 2015), there was a high degree of concentration for such a person being passionate about her work and curious. In 2012, the respondents considered that these people had an above-normal intelligence, unlike in 2015, when this characteristic was not as relevant.

On the other hand, the people surveyed in the two years said that some of the reasons why a person decides to work in science include following a vocation for research and having interest in knowing the truth and solving problems of people.

The respondents in 2012 and 2015 considered that science would bring more benefits than risks. Generally, people thought that advances in science and technology could bring many benefits, but these benefits would be limited. At the same time, over a longer period of observation, some risks to health and environment could be reflected.

As for the question on the problems that innovation could solve, the respondents in 2012 and 2015 considered that innovation could solve problems of telecommunications and health, following in importance the performance of companies (see Fig. 11.3).

For the attitudes and assessment category, the results showed that the spaces of interest for respondents in science and technology were expressed in the media, again highlighting the role of abstract media science (manual science). In addition, respondents' perceptions emphasized that science

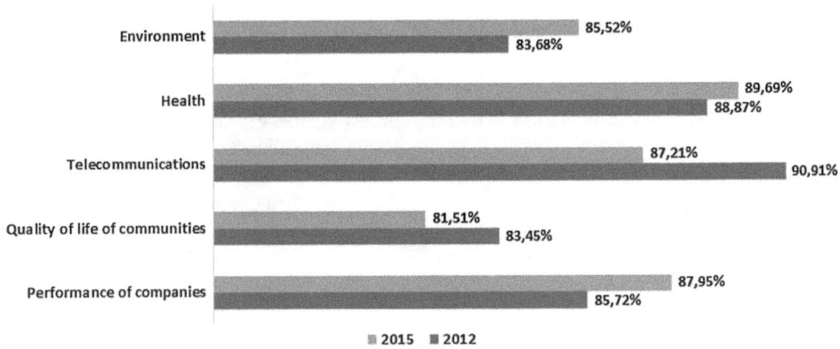

Fig. 11.3 Problems that could be solved with innovation. Source: SPPS&T, 2012 and 2015 OCyT

brings more benefits than risks. However, respondents said that the risks become latent over the long term in terms of health and the environment. Moreover, the knowledge generated was useful in different aspects of society; however, the people surveyed related S&T more to health and environmental aspects.

11.3.4 Social Appropriation of Science and Technology

In Colombia, the national policy and strategy for the SAS&T were defined as "activities that contribute to the consolidation of a scientific and technological culture in the country, that is, the scientific and technological community, who make decisions about science and technology, the media and the general public and to develop a greater capacity for critical analysis of science, technology and its relations with society and nature" (DNP, 2009, p. 27).

Considering this definition, the percentage of people who had a job related to science and technology was only 19.17% in 2012 and 16.62% in 2015. This situation may be because the respondents did not associate their work with scientific and technological issues. For the participation of people in spaces that promoted science and technology in the country, 47.27% of the respondents in 2015 visited natural parks, a trend similar to that observed in 2012 (see Fig. 11.4).

The guidelines established at the international level regarding the realization of indicators that account for aspects related to health and the

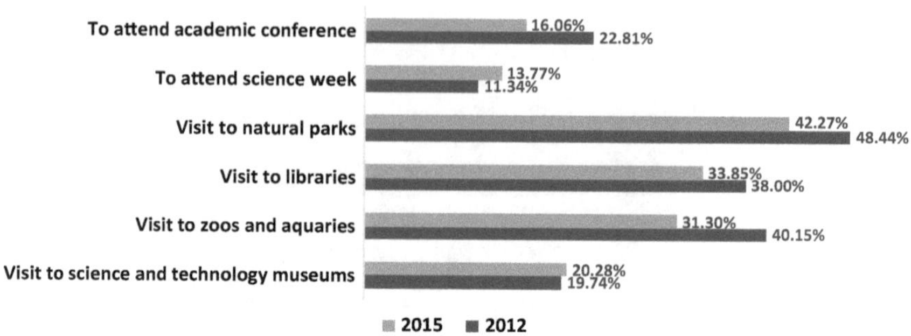

Fig. 11.4 Activities related to science and technology done by respondents. Source: SPPS&T, 2012 and 2015 OCyT

attitudes of people in relation to the cure of diseases include questions such as what have you done or what would you do if you had a disease such as hypertension/cancer? Regarding hypertension, people considered that it was important to follow the doctor's treatment and to ask the opinion of more than one doctor. For the two years of analysis presented here, a similar trend was observed. For cancer, the trends presented were similar: people preferred to follow the doctor's medical treatment, and many of them would also seek more than one doctor.

The main results for the SAS&T dimension showed that although the respondents in the survey years (2012 and 2015) claimed to visit natural parks, look at products' expiration dates when buying processed foods and follow the doctors' treatments when suffering from diseases such as cancer and/or hypertension, the SAS&T activities were not completely framed in terms that promote or consolidate the country's scientific and technological culture. However, this may have been due to the kind of questions asked, which require comparability with surveys conducted in other countries.

11.3.5 Citizen Participation

According to Ocariz, C. J. C., & Masi, S. D. (2016, p. 207), "Citizen participation in S&T is itself an action for the democratization of science and technology. With the various mechanisms of participation what is seeks to gradually incorporate citizens into the functioning of a national S&T system".

Participation mechanisms related to problems that concern the community showed that people's participation was very low, even when situations directly affected them. This result was also due to a culture of letting others do for oneself. For the respondent lending support to a referendum via his signature, in 2012, 26.63% confirmed their support, while in 2015, this percentage was 8.91%.

For the question "If you are invited to a meeting to make decisions on the impacts of science and technology would you attend?", for 2012 and 2015, 81.33% and 70.34%, respectively, expressed their willingness to attend a meeting related to S&T decision making.

In terms of citizen participation, the rate was low in the different activities presented to the respondents, which was more evident in 2015 than in 2012. This result was paradoxical in relation to what the respondents said in both years when asked if they would participate in a decision-making meeting on the impacts of S&T. In that case, the surveys found that the majority of the respondents would be willing to participate in such a meeting.

11.3.6 Public Policies

S&T policies are linked to institutional recognition. In addition, they are associated with the democratization of scientific and technological developments and with control over impacts and research results. In this sense, in terms of SPPS&Ts conducted in 2012 and 2015, it was found that the policy is associated with the proximity of the respondents to these issues.

For this category, related questions were asked about how S&T could help solve the country's political and social problems in the country. A large number of people in both surveys (see Fig. 11.5) believed that scientific and technological research could help solve epidemic problems (52.23%) and environmental pollution (53.89%).

In 2012, respondents considered that ethical committees and medical associations were the entities that should regulate scientific developments, while in 2015, most people believed that the government and universities should do so. Moreover, few respondents recognized Colciencias: in 2012 only 21.58% knew what the entity was, whereas in 2015, only 38.01% of the respondents recognized the institution.

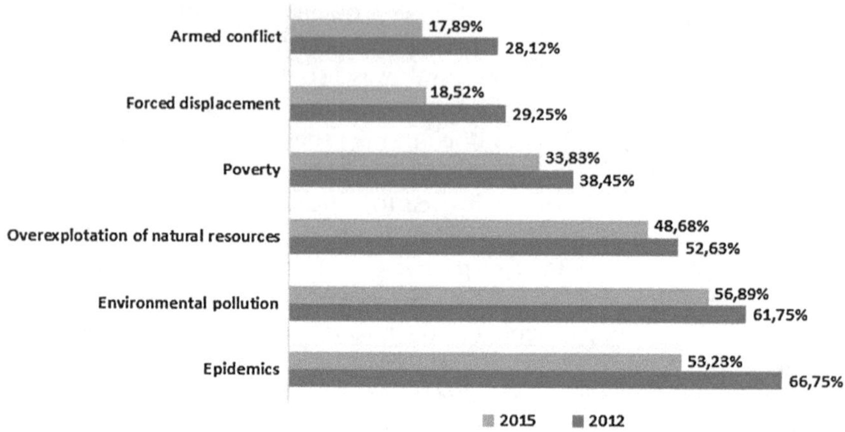

Fig. 11.5 Solutions generated by science and technology according to respondents in both surveys. Source: SPPS&T, 2012 and 2015 OCyT

11.3.7 Perception Indexes

From the survey results, 12 indexes were calculated. These indexes were related to citizens' perceptions about science and technology and seek to analyze different perspectives regarding the use of information media to learn about science, technology and information on subjects of interest related to medicine, health, environment, education in science and scientific vocation, environment and ecology, among others. Table 11.2 shows a comparative analysis of the indexes for 2012 and 2015.

Science, technology and innovation perception index. The trends show that the proportion of the Colombian population who consider science and technology in daily life increased in 2012. The science, technology and innovation perception rate shows that people assigned a 51.13% importance to the aspects related to the SAS&T, considering the high (41.03%) and very high (10.10%) importance that were assigned. In the 2015 version, the index presented results in which the majority of Colombians tended to assign an importance of 87.62%, including the variables high and very high (56.44% and 31.18%, respectively).

Media index in relation to science. This index seeks to determine the relationship between media and the diffusion of science, by inquiring about people's media use and consumption to learn about science. In 2012 and 2015, the groups with the highest percentages were 1, 2 and 3

Table 11.2 Comparative results of the 12-perception index calculated from the perception surveys of the years 2012 and 2015 at the national level

Science, technology and innovation perception index

Year	*Nothing important*	*Less important*	*Not much, not little*	*Important*	*Very important*
2012	2.31%	0.86%	0.31%	5.62%	90.89%
2015	5.32%	2.80%	8.16%	11.24%	72.48%

Index of media in relation to science

Year	Group 1	Group 2	Group 3	Group 4	Group 5	Group 6	Group 7	Group 8	Group 9
2012	19.43%	27.86%	25.59%	8.81%	2.35%	2.87%	7.99%	1.92%	3.18%
2015	32.94%	15.57%	25.40%	9.04%	2.64%	2.77%	6.30%	2.24%	3.11%

Index of media in relation to technology

Year	Group 1	Group 2	Group 3	Group 4	Group 5	Group 6	Group 7	Group 8	Group 9
2012	9.57%	2.77%	2.00%	2.82%	2.03%	20.77%	27.14%	25.48%	7.41%
2015	6.84%	2.01%	1.57%	1.69%	2.18%	20.68%	12.54%	44.13%	8.36%

Index of space use for science communication

Year	Very low	Low	Not low, not high	High	Very high
2012	48.72%	19.86%	13.36%	10.63%	7.42%
2015	52.87%	14.08%	20.39%	6.89%	5.77%

Index of appropriation of science and technology in everyday life

Year	Very low	Low	Not low, not high	High	Very high
2012	0.80%	8.89%	38.54%	40.34%	11.43%
2015	0.36%	0.60%	11.42%	56.44%	31.18%

Information index in relation to medicine and health

Year	Group 1	Group 2	Group 3	Group 4	Group 5	Group 6
2012	11.02%	37.21%	16.34%	0.92%	34.28%	0.23%
2015	12.08%	31.89%	15.81%	0.26%	39.78%	0.18%

Index of information in relation to environment and ecology

Year	Group 1	Group 2	Group 3	Group 4
2012	59.03%	9.72%	30.80%	0.45%
2015	68.83%	12.25%	18.39%	0.53%

Index of interest in science, technology, environment and health

Year	Group 1	Group 2
2012	65.02%	34.98%
2015	73.27%	26.73%

(*continued*)

Table 11.2 (continued)

Science, technology and innovation perception index					
Year	*Nothing important*	*Less important*	*Not much, not little*	*Important*	*Very important*
Participation rate					
Year	Very low	Low	Not low, not high	High	Very high
2012	0.10%	17.66%	0.92%	3.12%	78.20%
2015	0.42%	0.96%	29.46%	1.05%	68.10%
Index of interest and utility of science, technology and innovation					
Year	Very little	Little	Not little, not much	High	Very high
2012	1.09%	0.65%	1.60%	14.28%	82.37%
2015	1.07%	1.27%	1.00%	2.38%	94.29%
Index of perceptions about science education					
Year	Group 1	Group 2	Group 3	Group 4	Group 5
2012	66.88%	4.52%	24.95%	1.06%	2.59%
2015	51.41%	17.98%	8.16%	20.90%	1.54%
Index of perception about the scientific vocation					
Year	Group 1	Group 2	Group 3	Group 4	Group 5
2012	37.09%	18.56%	11.96%	21.83%	10.55%
2015	46.41%	28.02%	9.68%	9.06%	6.83%

Source: From the results of the 2012 and 2015 SPPS&Ts by the OCyT.

(see table with group descriptions). In 2012, group 2 represented the highest percentage, indicating that people use and consume television and radio the most to stay informed about science. In 2015, the highest percentage was group 1, indicating that the internet is the most used means by which to learn about science and the most consumed media are the internet and television. Group 3 remains similar in both studies, indicating that television is the preferred medium and that the most consumed venues are TV and the internet.

Media index in relation to technology. This index seeks to determine the relationship between media and the diffusion of technology, from investigating people's use and consumption of media to learn about technology. The groups that predominated in both surveys were 6, 7 and 8. For group 6, the results were similar, indicating that approximately 20% of the population is informed about technology through television and the means of greater consumption are television and internet. In the 2015 survey, there was a decrease in the percentage of group 7s participation, which may show that television is still the preferred medium, while radio has decreased in preference as a means of communication for technological information. In group 8 in 2015, there was an increase of approximately 20%, again evidencing the internet's positioning to inform about scientific-technological issues.

Space use index for science communication. This index analyzes and determines the frequency with which the population attends activities and institutional spaces of communication in science and technology for broad audiences. The 2012 results showed a higher percentage in the very low and low categories, while for 2015, the categories of higher incidence were very low and neither low nor high, showing a certain increase in the number of citizens surveyed participating in events related to visits to museums, botanical gardens and libraries, among others.

Index of appropriation of science and technology in everyday life. This indicator reviews some associated practices, such as the exploration of labels, instructions for use, product contraindications and environmental issues, which seek to analyze the links between science, technology and society. The results show that in 2012, most of the population was concentrated in the index scale related to the category neither low nor high (approximately 80%), while in the year 2015 more than 80% of the respondents were in the high and very high scales. This result evidenced a great change in the trend in which the population appropriates science and technology in their daily lives with regard to more informed purchasing decisions when reviewing, for example, labels, instruction manuals and contraindications.

Information index in relation to medicine and health. This index determines how well informed people are and their media preferences for health issues. The groups with the greatest representation in the two surveys were 2, 3 and 5. Group 2 had a higher incidence in the 2012 survey, whereas in 2015, group 5 predominated. This result showed that people are neither informed nor manifest interest in medicine and health. In both measurements, less than 40% of respondents were informed or entertained by medical and health topics.

Index of information on environment and ecology. This index was developed to determine citizens' interest with respect to environmental and ecological issues by establishing the level of information and/or entertainment. The two survey results agreed that the largest number of people were in groups 1 and 3 (low interest). However, in 2015, approximately 80% were concentrated in groups 1 and 2, which meant that interest in information on the environment and ecology has been increasing, mainly due to the dissemination of information on experimental changes with regard to the environment by climate change, higher pollution levels and current environmental risks. These changes have motivated the population to learn about that environmental conditions in which they live and the measures that can be taken to care for the environment and natural resources.

Index of interest in science, technology, environment and health. This index seeks to group the previous themes regarding respondents' use of means for learning about science, technology, environment and health or to entertain themselves with related topics. In both surveys, group 1 predominated, with an increase in participation in the survey in 2015, which showed that people used the media for entertainment in all four subjects and had a special interest in science and technology. These results showed that, increasingly, the population incorporates into their daily lives habits regarding the use of means that affect the processes of appropriation of subjects related to health, environment, science and technology. In the last themes reviewed, their characteristics and dynamics can be explained through science and technology.

Participation rate. This index seeks to determine the frequency with which people participate or are willing to participate actively in issues related to science and technology. In both surveys, responses were concentrated on the very high scale, with the highest proportion in 2012 (78.20%), while in 2015, there was a slight decrease (68.10%). These results showed that the population that wants to participate in science and technology has remained interested using the mechanisms of citizen participation, attending events where they can give their opinion and taking actions in situations that could affect their quality of life while having information on science and technology.

Index of interest and utility of science, technology and innovation. This index seeks to determine how respondents consider science, technology and innovation to be useful and interesting for everyday life. In both surveys, the predominant category was very high, with an increase from 82.37% to 94.29%. These results showed that the population increasingly perceives science, technology and innovation as useful and interesting for everyday life, which implies that new technological and scientific advances are increasingly impacting the daily lives of the entire population. These results showed a greater participation of the media as generators of opinion in these subjects, motivating the population to investigate more and showing the importance of science, technology and innovation as conducive to change and development in the country.

Perception index on science education. This index seeks to determine the relationship between the values inculcated to people in childhood and the value these have on science education. In the 2012 survey, groups 1 and 3 predominated, whereas in the 2015 survey, groups 1 and 4 predominated. In both surveys, the highest percentage group was 1, indicating that people agreed that the school worked on useful themes and instilled values related to science and technology. These results highlighted the importance of education and family in early childhood to motivate science's importance. For this reason, it is important that Colombian education foster similar habits of science such as creativity, curiosity and critical thinking, which would allow the population to become more interested in science and technology subjects and show its importance in daily life. These themes are the engine of development and progress in closing knowledge and information gaps.

Index of perception about the scientific vocation. This index seeks to characterize people by their scientific vocation inquiring about the interests of educational training directed at programs that have a relationship and/or contribute to scientific and technological construction. In the 2012 survey, the results were concentrated in groups 1 and 2, whereas in the 2015 survey, the answers in groups 1 and 4 predominated. However, in both surveys, group 1 had the highest percentage, indicating that people preferred their children to study medicine or engineering. In addition, common perceptions were identified, such as engineering is the best paying job, studying social sciences makes a job more difficult to find in Colombia, studying basic sciences can only lead to a career devoted to teaching, and the most important thing when choosing a profession is the future promise of a good salary. This group also believed that scientists are solitary and rigorous and that people decide to work in science to earn money, because of a research vocation, because it offers prestige, to solve people's problems, to have power, to know the truth and to understand the environment.

The results of perception indexes and trends are fundamental to analyzing and understanding how the Colombian population appropriates S&T every day, how different public and private strategies empower society using S&T and where mass media demonstrate key themes for the individual and society. Policymakers can use these results to formulate strategies and programs that allow a higher appropriation of scientific and technological culture in Colombia, which results in more development, more responsible decision making and equality in knowledge access.

11.4 Conclusions

The results of the surveys of perception for the years 2012 and 2015 show telling results in terms of the perception of Colombians in the subjects of S&T. Some aspects of the surveys are highlighted below: (1) National television and social networks are the media preferred by the people surveyed. This same tendency is found in the media by which people prefer to learn about science and technology; (2) There is a high level of interest in S&T. The media and the university have aroused the interest of the respondents on these issues; (3) Scientific and technological knowledge is useful in the collective imagination of the people surveyed, in health care and prevention of diseases and to understand the world. It is somewhat useful for the formation of their political and social opinions; (4) For people's practices when buying processed foods, it is found that an expiration date is always provided. The foods are less frequently found to contain no substances harmful to health, and it is almost never seen that their content can affect the environment; (5) For diseases such as hypertension and cancer, people prefer to go to a health professional and learn about their illness; and (6) In the collective imagination, people still believe that science and technology are performed, but the percentage that supported this statement in 2012 is lower.

Respondents perceived that education was linked to greater future opportunities. They considered it important to learn systems at school for the future and to choose a profession that represents a good salary. They also thought that engineering was a better paid job.

The perception surveys conducted between 2012 and 2015 indicate that actions should be implemented to increase the population's interest in S&T. However, the mechanisms or the main factors implementing these initiatives should be considered. Therefore, providing complete information is vital to empower the citizen as a decision-making agent, who can contribute from an active role in the field of science, technology and research. Notwithstanding, some consider, "It is a fact that citizens demand the opening of channels of expression and participation although, at the same time, this does not entail the exhaustion of technocratic forms of public affairs management or the withdrawal of support for expert systems as a source of legitimization of policies" (Polino & Chiappe, 2010, p. 113).

In addition, the media dynamics by which people now inform themselves must be understood to socialize the information sufficiently and in a way that will achieve diffusion and effectiveness. In this order of ideas, the message to be expressed and disseminated must also be considered because science and technology are being developed in Colombia for the target population to be more economically developed and less dependent on other countries.

Another important aspect is new courses of action that involve knowing the target population, since it is necessary to foster more effectively critical thinking and curiosity. These factors are vital for science, technology and innovation and have not been the main characteristics of aging in Colombia.

The strategy that Colciencias wants to implement toward support for education must be defined. Respondents felt this issue should receive the highest budget and felt that they were more informed about it. This positioning in the citizen's agenda implies a special level of attention to provide information and meet expectations about the relationship between education, science, technology and innovation. The previously explicit perception coincides with the interventionist tendency that emerged after the seventies, in which science was channeled over more interventionist policies; in which the incidence of technical, administrative and legislative instruments becomes evident; and public participation becomes a constant in institutional initiatives to intervene in science and technology (Palacios et al., 2005, p. 5, sf).

In summary, the results of both surveys show similar trends indicating that Colombian citizens have preferences for science communication by space, allowing advancements in the SAS&T process in daily life. Other results indicate a higher visibility of activities such as reading, a proactive search for technical and scientific data on diverse products and services used every day, a shift in consumption and lifestyle to counter environmental problems and participation in collective actions that seek to generate changes in the society-nature relationship.

Moreover, perception indexes and trends are fundamental for analyzing and understanding how the Colombian population appropriates S&T every day, how different public and private strategies have achieved an empowering society using S&T, and where the mass media have been important in demonstrating these themes to both the individual and society. Policymakers can use these results to formulate strategies and programs that allow a higher appropriation of scientific and technological culture in Colombia, which can result in more development, more responsible decision making and equality in knowledge access.

Finally, universities and government are called upon to take a more active role in the S&T field to have tangible results for citizens not only in basic science areas but also to foster innovation in areas such as S&T in social settings and in new agendas such as post-conflict. Thus, they can nourish the collective imagination that today dominates these disciplines and give S&T a greater reach.

Acknowledgments This document was prepared based on the Final Report of the IV National Survey of Public Perception of the ITC delivered by the Observatorio Colombiano de Ciencia y Tecnología to Colciencias in the framework of the Consultancy Services Contract No.777 signed between Colciencias and the Observatorio Colombiano de Ciencia y Tecnología on September 24, 2015. Research for this study was funded by the United Nations Development Program—UNDP. We gratefully acknowledge the research team that helped conduct this work and the comments by anonymous reviewers to improve the manuscript.

References

Daza-Caicedo, S., Lozano-Borda, M., Bueno Castellanos, E. M., Gómez-Morales, Y. J., Salazar Acosta, M., Jaime, A., ... & Pérez-Bustos, T. (2014). Percepciones de las ciencias y las tecnologías en Colombia. Resultados de la III Encuesta Nacional de Percepción Pública de la Ciencia y la Tecnología.

Daza-Caicedo, S., Paz, J. C., Camelo, C. R., Bravo, D. H., Barrera, E. A. R., & Hernández, I. N. C. (2009). *Percepciones sobre la ciencia y la tecnología en Bogotá* (S. D. Caicedo, Ed.). Observatorio Colombiano de Ciencia y Tecnología.

DNP. (2009). *Política Nacional de Ciencia, Tecnología e Innovación.* Documento Conpes 3582. Bogotá.

Ocariz, C. J. C., & Masi, S. D. (2016). Participación Ciudadana en las Política de Ciencia y Tecnología. Estudio Comparativo de los Marcos Normativos en Ciencia y Tecnología de Bolivia, Colombia, España y Paraguay. *Revista Internacional de Investigación en Ciencias Sociales, 12*(2), 205–222.

Palacios, E. M. G., Galbarte, J. C. G., Cerezo, J. A. L., Luján, J. L., Gordillo, M. M., Osorio, C., & Valdés, C. (2005). *Ciencia, tecnología y sociedad: una aproximación conceptual.* Organización de Estados Iberoamericanos (OEI).

Polino, C., & Chiappe, D. (2010). *Percepción social de la ciencia y la tecnología. Indicadores de actitudes acerca del riesgo y la participación ciudadana.* El Estado de la Ciencia.

CHAPTER 12

Innovation, Entrepreneurship, and Technology-Based Clusters in Emerging Economies: How Can the Madrid Aerospace Cluster Reinforce the Mexican Aerospace Industry?

José Manuel Saiz-Alvarez and José Ramón Gutiérrez-Martín

12.1 Introduction

Innovation processes driven by technology-based methods and entrepreneurship are of particular importance in industrial clusters and are mainly defined by the combination of regional foreign direct investment (Behera, 2017) and environmental-related factors (Bellantuono, Carbonara, & Pontrandolfo, 2017). As a result, when industrial clusters are numerous

J. M. Saiz-Alvarez (✉)
EGADE Business School-Tecnológico de Monterrey, Guadalajara, Mexico
e-mail: jmsaiz@tec.mx

J. R. Gutiérrez-Martín
ESDEN Business School, Madrid, Spain
e-mail: jgutierrez@esden.es

C. I. Pardo Martínez et al. (eds.), *Analysis of Science, Technology, and Innovation in Emerging Economies*,
https://doi.org/10.1007/978-3-030-13578-2_12

and active in the country, nations achieve sustained and sustainable business growth. In this process, the combination of knowledge, information and communication technologies (ICTs), and industrial processes play a vital role to form the Industry 4.0.

Depending on the structure of the global value chain used in the corporation, clusters divide knowledge into intra-cluster knowledge sharing ('local buzz') and externally sourced knowledge ('global knowledge pipelines') (Guimón & Paraskevopoulou, 2017). When the country starts from low levels of economic growth and development, it is desirable for the cluster to apply both types of knowledge simultaneously to accelerate business growth. Also, one of the clusters with more development and value is the aerospace industry located on both sides of the Atlantic, where the regions of Querétaro (Mexico) and Madrid (Spain) have robust international importance.

The novelty and contributions of our book chapter deal with the science and technology applied in both regions to propose ideas to create a 'Geometry from Diffusion of Innovation' in emerging economies, particularly in Mexico, to create a European-American triangle of sustainable cluster growth. A geometry of innovation focused on economic progress and was formed by the conjunction of growth triangles, trans-border regions, nodes, poles of growth, and axes/corridors. As a result, the development of the state of Querétaro in Mexico, combined with the state of Jalisco, also known as the Mexican Silicon Valley, can accelerate the formation of a Mexican axis of growth between the states of Jalisco, Guanajuato (as a bridge), and Querétaro, to lead a harmonious development of the territory (Fig. 12.1). In the figure cities in black (Querétaro (QRO), Seattle, Montreal, Wichita, and Toulouse) already form the axis of growth, while the proposal is to integrate Guanajuato (GTO), Guadalajara (GDL)-Jalisco and Madrid in the cluster to create a double pivot of economic growth. An industry where the convergence of entrepreneurship-based sectors working in a triple helix (universities-government-private sector) frame, civil society, and technology-rooted clusters provides models focused on regional innovation (Etzkowitz, 2014) and economic growth.

The innovation process in industry-based clusters is fundamental for companies to create competitive groups defined by leadership in their respective sectors, and innovation must be continuous over time to have medium- and long-term effects. However, when firms refuse to innovate, they will tend to disappear in the long-term, as they cannot be notorious in the market.

In accordance with the small-scale technology theory developed by Wells (1983), the enterprises in small-scale markets like firms existing in

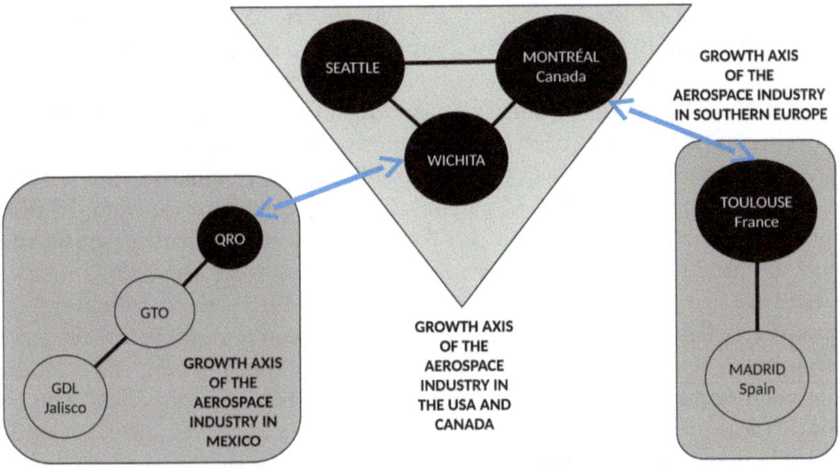

Fig. 12.1 Growth axis in the aerospace industry. Source: Authors

developing and emerging countries are featured with higher flexibility and lower technology costs, which enable them to capture the market with a faster speed and a lower price (Pu & Zheng, 2015). When this higher flexibility is linked to innovation processes, firms generate first-order competitive advantages, defined by research and development (R&D) and innovation, which makes them very difficult to be imitated by competitors. This fact creates sustainable benefits over time, which leads to higher financial strength in the organization. An advantage that favors companies to reach sustained leadership positions over time in the sector where they operate. At this respect, Johanson and Vahlne (1977) propose an international enterprise network model, which indicates that the internationalization process of companies is a cycle defined by the use of knowledge, and the application of resources to establish, develop, maintain, and finish the process of network relationships.

The objective of our work is to analyze how the Madrid Aerospace Cluster (Spain) (Martí Fluxá, 2005) can reinforce the growth axis developed in Querétaro Aerospace Valley linked to the high-tech structure established in the state of Jalisco (Mexico) to create an international axis of economic growth, as shown in Fig. 12.1. In fact, Jalisco is known as the Silicon Valley of Mexico, endowed with the most significant technological hub in Latin America formed by 460 startups and more than 6000 technology-based firms, mainly located in the Metropolitan Area of Guadalajara (Jalisco, Mexico). Complementary to this industrial hub in the state of Jalisco, the state of Querétaro is impulsing a new aerospace cluster to develop this industry in Mexico. In this sense, entrepreneurs working in technology-based groups endowed with innovation can create firms defined in the Industry 4.0 to build aerospace clusters that, in the case of the region of Madrid (Spain), are active with more than 15,000 industrial firms located in the territory (Cotec, 2017). In this sense, the authors have the objective of extracting new ideas from the Spanish experience to strengthen the Mexican aerospace industry to foster the inter-Atlantic collaboration between Mexico and Spain. To achieve this goal, first, the development of industrial clusters is analyzed. Then, the relation between technological innovation and endogenous growth, the role of innovation processes in sectoral clusters, and the cluster location are studied. Finally, the Case of the INNOVA Model of Madrid Aerospace Cluster and its validation are both analyzed.

12.2 Literature Review

12.2.1 The Importance of Entrepreneurship-Based Innovation

The combination of private entrepreneurship and public intervention fosters the creation of clusters and economic poles that are beneficial to countries and societies. Entrepreneurs are a fundamental part of the economic structure of nations. Business owners are generally conscious of their role in their communities giving rise to their feelings. Because the resulting socioemotional wealth is intended to be enjoyed by the entrepreneur alone, this view reflects a mostly self-interested approach to decision-making (Newbert & Craig, 2017). This emotional decision-making links to business decisions rooted in corporate social responsibility (CSR) to help change societies. A business strategy that is having an increasing role to play, mainly in developing and emerging countries.

Although Bowen (1953) is considered the modern father of the CSR, some other authors (Bernard, 1938; Kreps, 1940) have seminally thought about this issue (Milian, 2015). These seminal works have been enlarged by Davis (1960, 1967), cited by Schwartz and Carroll (2003), in which he asks what the entrepreneur owes to society (social mortgage) (Ramírez, 2016; Saiz-Álvarez, 2017) and what responsibility companies have in front of the community. The role of the entrepreneur in almost all the entrepreneurial ecosystems of the planet is fundamental. These ecosystems create triple helix models (Bergek, Jacobsson, Carlsson, Lindmarki, & Rickne, 2005; Carlsson, Jacobsson, Holmén, & Rickne, 2002; Carlsson & Stankiewicz, 1991; Edquist, 2005) defined by their diffuseness, heterogeneity, intense focus on institutions, low visibility of the role of individuals in the innovation process, and system boundaries (Ranga & Etzkowitz, 2013). Triple helix models are recently drifting toward quadruple helix models where non-governmental organizations (NGOs) have a vital role to play. As a result, quadruple helix models are permeating both civil societies and organizations due to the need for greater awareness and social support toward the most disadvantaged and the population at risk of social exclusion. Quadruple helix models reinforced with positive crowding-in effects resulted from the interaction among universities, NGOs focused on social assistance and the fight against poverty and inequality; business organizations, with the creation of positive externalities regarding job creation and wealth for a good part of society; and the public sector, mainly by applying fiscal policy. Changing process based on the 7-K (know-how, know-where, know-when, know-whose, know-why, know-who, and know-what) (Saiz-Alvarez & García-Ochoa, 2008) in order to have success.

When entrepreneurs link their business strategies on CSR, they lay the foundations to achieve business leadership that must be sustainable over time if they want to make long-term positive results. Leadership in family businesses is an essential aspect to have in mind when structural change is carried being out in firms aiming business success. Being a leader with social responsibility is an obligation in a globalized world defined by the growing competition in regional blocks located in different regions of the planet. CSR is part of our lives to improve corporate's goodwill, especially in the case of family firms. In this sense, the combination of quadruple helix schemes, formed by the combination of universities, the public sector, businesses, and NGOs, generates social and economic wealth.

In this respect, Latin American and the Caribbean (LAC) countries are strengthening quadruple helix models for social change based on leadership. A leader is active which is defined by ten personality traits: honesty, capacity to delegate, communication, confidence, commitment, creativity, intuition, inspiration, positive attitude, and openness, especially toward his/her team (Prive, 2012). While in developed nations individual entrepreneurship is the predominant one, in developing countries it would be more desirable to create a social enterprise to integrate the most disadvantaged communities, as well as to achieve a process of sustainable social change. Thanks to social entrepreneurship, combined with CSR-based business policies, the methods of social transformation lead to the achievement of solidarity-based societies defined by the creation of a broad middle class that sustains the State with tax payment. For this reason, public intervention in developing countries is more necessary due to the crowding-in effects generated, since the private sector is fragile. In this sense, the model existing in the first-world countries must necessarily be different from the model designed for the developing world, if a real socioeconomic change is desired. This is why the public intervention has been very intense during the creation of industrial clusters in Mexico to benefit the Mexican aerospace industry.

12.2.2 *The Development of Industrial Clusters*

Every cost focused on reducing communication-related expenses, and every new facility for the free interchanging of ideas between distant places alters the localization of industries. Lowering tariffs and freights in the transportation of goods tends to be concentrated in particular sectors and locations, especially when transportation costs and freights are very high. The importance of industrial concentration and the emergence of location rents have been studied mainly in the 1990s, with the renewed interest in economic geography (Krugman, 1991a, 1991b). At the beginning of the twenty-first century, however, the territory starts playing a crucial role in the growth sustainability, which can be endogenized and referred to as the 'social capital' of the region (Vázquez-Barquero, 1999) that is directly influenced by culture (Crociata, Furia, & Agovino, 2018).

The economies of concentration lead to gains in efficiency and productivity, derived from the focus of companies in a territory. In this sense, different authors have subsequently highlighted how these economies emerge, mainly (1) by generating a 'collective spirit' (Perroux, 1955); (2) because there are relations of mutual trust between the stakeholders of the concentration (Saxenian, 1994); and (3) by developing a relatively homogeneous system of values and thinking (Becattini, 1979). As a result, collective efforts in business can be achieved, which enhances productivity, cash flow and EBITDA generation, and further business collaboration.

Moreover, Becattini (1989) defines industrial districts as 'socio-territorial entities, characterized by the active presence of a community of people and companies in a given geographic and historical space, where there is perfect osmosis between the local community and businesses'

(p. 23). Also, Saxenian (1994), making a biological analogy, affirms that industrial concentrations are 'protein sites' in which firms achieve 'speed advantages,' as there are more natural access to markets, more demanding consumers, abundance of skilled labor, dense supplier networks, industries and auxiliary services, rivalry among companies in the sector, local and regional connections favoring innovation, and flexible diffusion of technologies. These results reinforce in the case of the existence of substantial economies of agglomeration and good transport infrastructures and facilities (García-López & Muñiz, 2010).

Besides, Cossentino, Pyke, and Sengenberger (1996), referring to industrial districts, believe that their common denominator is their high dynamism summarized in achieving more significant levels of efficiency and dexterity in production through specialization. As a result, there is faster and easier access to the full range of the industrial value chain expertise, due to the availability of local talent and capabilities and the presence of an industrial atmosphere that allows rapid diffusion of new ideas. Finally, the combination of the existence of a mix of cooperative and competitive relations ('coopetition') between companies and the promotion of entrepreneurial culture results in the development of competitive advantages, servitization[1] included (Sharma & Singh, 2017).

Similarly, Dini (1996) points out three competitive advantages found in industrial districts: (1) *concentration economies*, which attract both suppliers of inputs and specialized professionals making possible to enhance the supply of these factors by increasing the efficiency in their respective markets. As a result, operating costs will reduce quickly; (2) *advantages of specialization* to achieve economies of scale and reduction in inventories thanks to efficient learning processes resulted from the simplification of procedures, and (3) *benefits of integration*, which enables the exchange of experiences, the expansion of networks, and the joint development of strategic knowledge. Consequently, it improves the technological diffusion within the district and reduces the level of uncertainty of entrepreneurs in the strategic decision-making process.

Regarding the types of industrial districts, Markusen (2002) indicates four categories, as follows

[1] Coined by Vandermerwe and Rada (1988), the servitization of products deals with the strategy of creating value by adding services to products or even replacing a product with a service (Baines, Lightfoot, Benedettini, & Kay, 2009), as happens in the industrial clusters.

a. The *Marshallian Industrial District*, where there are dragging of suppliers and concentration around their clients (Becattini, 2017);
b. The *Hub and Spoke District*, where small and medium enterprises (SMEs) focus on large enterprises, as in the case of the Seattle Region (Gray, 1996);
c. The districts constituted by industrial subsidiaries of the significant non-local companies, forming networks with SMEs and large local businesses, and
d. Districts built around public or private institutions, such as technology centers, sector associations, and sector or local policy organizations.

Industrial districts generate economies of location to become centers of growth. The French economist François Perroux (1955, 1960) initially developed the concept of 'Growth Pole,' as the area where surge polarization effects or spatial divergences of income. In other words, the poles are geographic areas where there is greater economic growth, which in turn spatially expands to be connected to other poles to form the so-called Growth Axes or Corridors. Besides, Perroux affirms that the pole formation depends on the interaction of increasing returns to scale, lower transaction costs, and significant markets capable of absorbing the production generated by the growth pole. These factors reinforce and self-sustain the geographic concentration of firms to emerge a process of 'cumulative circular causation' (Kaldor, 1963; Myrdal, 1957) and perpetuate divergences between regions. Therefore, they find effects of absorption of economic activity and employment by the wealthiest areas. Results are more significant than those results generated from the diffusion to the less favored regions.

In this sense, Krugman (1991a) developed some years later the center-periphery model, which explains that economic concentrations generate centripetal forces (external economies) and centrifugal forces (external de-economies) to reach a saturation point in the concentration. In addition, Krugman, Fujita and Venables (2000) add that economies achieve multiple equilibria, and therefore, the equilibrium stability must be analyzed. Finally, transaction costs increase more than proportionately because of the strategic behavior of the groups of losers in the negotiation, who behave as lobbies or 'Olsonian' groups with low organizational costs.

12.2.3 Technological Innovation and Endogenous Development

As an evolution of ideas around industrial clusters, Castells and Hall (1994) developed the concept of means of technological innovation (MTI). This approach emphasizes the globalization processes applied in national and regional economies under the technological paradigm of ICTs by following an analysis of high-tech industries' concentrations in several countries, which they called 'Technopolis.' As a result, these authors define MTI as 'the particular set of relations of production and management based on a social organization, which mostly shares a professional culture and instrumental objectives, aimed at generating new knowledge, new processes and new products' (p. 23). Moreover, these authors highlight the organizational transformation from pyramidal structures to more flat and horizontal ones, due to new, more flexible and specialized production capacities. In this sense, companies face new conditions for competition, allowing the emergence of very dynamic and brand new industrial areas with a high capacity for technological innovation and productive specialization, not only in ICTs, but also in other groups of technologies related to biotechnology, new materials, and aerospace industries.

The fundamental characteristic of these innovative spaces is their ability to create synergies to raise the final production value by combining their elements in the territory to maximize production. Consequently, Castells and Hall (1994) classify technological spaces into

a. *Industrial complexes of technological innovation*, formed by industrial concentrations to contain centers of science and technology research to generate new knowledge, products, and industrial processes located in the agglomeration.
b. *Scientific cities*, defined as concentrations of public and private research centers, developed with the objective of generating the critical mass needed to advance scientific knowledge.
c. *Technological parks* conceived as industrial areas planned by the public administration, with the objective of locating and delivering services to high-tech companies.
d. *Metropolitan high-tech industries*, defined as large metropolitan areas located in industrialized countries, endowed with the highest concentrations of high-tech industries, such as Los Angeles, Paris, and Tokyo, and

e. *Regional poles of technologies*, defined as concentrations of industries and technological centers promoted by the public administration to stimulate local development.

In the beginning, the formation of an MTI often found a certain spontaneity; nevertheless, to reach its maturity requires the organized search of positive externalities derived from the concentration by the public and private agents. In this sense, and in a complementary way to the classification of industrial districts by Markusen (2002), Cooke and Morgan (1994) distinguish three models of spatial concentration of innovative activity:

a. *Leadership-based model*, where the public administrations organize the spatial level of research institutes and high-tech companies, setting up an infrastructure of scientific and technological parks, as in the cases of ZIRST areas of Grenoble (France) and Sophia Antipolis (France), with a population of more than 9000 citizens working in a 2300-hectare area.
b. *Spontaneous model*, where there is a natural and sustained level of high-tech development content, carried out by companies and R&D (research and development) and innovation centers located in the area. Examples of this model are the Californian Silicon Valley, the Oxford Area, and some European industrial districts.
c. *Network model* leading to the development of a local or regional system of innovation with a network composed of research institutes and firms (Cooke & Morgan, 1994). An example is the Baden-Wurttemberg innovation cluster.

Beyond the industrial concentration or cluster developed in the territory, its sustainability depends on the generation of a network and mutual learning between the stakeholders of the group. Besides, the concentration process derives into positive externalities, due to a higher division of labor. As there is not a single path of development, the formation of these clusters is conditioned by (1) the previous situation of the territory (industrial or agrarian); (2) networks between local partners; (3) the innovation and learning system of local environments; and (4) remittances from outside the area (especially from the public and private foreign investment) (Vázquez-Barquero, 1999). These factors allow us to explain several possible paths of gradual development, from the rural environment to an intermediate position called the 'Space of Innovation and Learning.' In this situation, it is possible to execute technology-based policies to encourage the development of technological poles, as an intermediate step to reach the 'milieu' model of excellence. This path is called the 'Two-Step Gradual Strategy,' where companies acquire knowledge to be organized by moving from the situation of a technology pole to a 'milieu' pole. In this process, internal resources of companies located in the territory gradually replace external resources from the State. An alternative strategy of 'Radical Change' also happens (Vázquez-Barquero, 1999).

In this sense, we have a clear example of this path in northern Italy (Emilia Romagna, Veneto, Tuscany, and Trentino). According to Rabelloti (1997), the evolution to a 'milieu' pole in the industrial regions of northern Italy is due to the combination of:

a. Spatially concentrated and sectoral-specialized SMEs to create clusters.
b. Social capital endowed with a robust and homogeneous cultural ground to develop standard codes of behavior (explicit and implicit).
c. Strong labor relations to exchange goods, services, information, and human resources.
d. Significant support networks developed by local public and private institutions.

At this respect, Cossentino et al. (1996) affirms that the industrial experience of northern Italy shows that the phenomenon of the industrial district can at least partly explain the high economic growth in developed economies after the Second World War. Also, the industrial areas located in northern Italy have demonstrated the ability to provide satisfactory social conditions rooted in a continuous dynamism, as part of their social capital, as a standard factor in many Italian industrial districts. A sustained vitality showed by

a. The predominance of SMEs, as in Tuscany, there is 1 company per 10 inhabitants; in Emilia Romagna, 87% of businesses have less than 11 workers, and in Veneto, there are 11 districts with firms that have from 4 to 20 employees.
b. Production is export-oriented, as in Veneto 5 of the 11 districts export 70% or more of the value of their production; in Emilia Romagna, the number of exports of packaging machinery increased

by 33%, and in the textile district of Prato the value of exports rose by 24%.

c. High-income generation, as in Tuscany the workers' income is 26% higher than the national average; the jewelry district of Belluno is around 92%, and in Veneto and Emilia Romagna, the unemployment rate is 6%, compared to 11% in the rest of Italy.

Besides, data from the European Cluster Observatory (2009) confirm that clusters are not created spontaneously in Europe, as are initially impulse by the State. In addition, this group configuration effort is more straightforward if there are previously particular conditions linked to competitive advantage and social capital in the territory (Sölvell, 2008). Consequently, a positive feedback loop phenomenon occurs, since clustering makes easier for regions to specialize and become more competitive in global environments. Besides, the ability to absorb and generate knowledge and innovation through a regional innovation system is of particular importance as the necessary precondition to achieving competitive advantages in the territory.

The Porter's 'Competitiveness Diamond' (1991, 2003) are collected to determine the competitiveness of a country or territory, and the capability to generate and absorb innovation. In fact, innovation plays a critical role and depends fundamentally on the regional innovation system, where private firms, public administration, R&D and innovation, infrastructures, and the environment interact. In addition, the performance of the innovation system primarily depends on technology, SME's networks, specialists in vital technologies (niche players), and the entrepreneurial culture. Besides, Porter (2003) states the possibility of anticipating innovation requirements, as a function of the nature of domestic demand, that is, the existence of the cluster of internal market forcing local companies to innovate. Finally, another critical competitive factor is how innovation varies along the sector's value chain, as it is possible to develop networks of companies from diverse sizes, different levels of scale efficiency, internationalization and vertical and horizontal integration, with close up sectors by technologies and products.

In fact, Porter (1991) explains that no nation could alone be competitive in all areas. On the contrary, each country owns competitive and non-competitive industries, and emphasizes the differences in the success of specific sectors, as competitiveness arises in particular locations. As a result, geographical proximity reinforces the strength of the diamond model.

In 2003, Porter abounded in this idea by explaining the importance of clusters in the regional economic performance of nations. Sölvell (2008) shows with data from the European Cluster Observatory that economic prosperity is related to the degree of strength of their clusters, and this relationship is also confirmed by reports, such as Competitive Regional Clusters (OECD, 2007) and Europe INNOVA (European Commission, 2008).

12.2.4 *Innovation Processes in Sectoral Clusters*

In the 1990s, sectoral clusters' goals mainly focused on maximizing external economies, location externalities, and social capital. Rabelloti (2003) and Sölvell (2008) group these externalities into economies of production and operation, and economies of innovation. In both types, the key factor to create externalities is knowledge overflow or knowledge spillover (Grossman & Helpman, 1991). In fact, the combination of these economies has a positive impact on cluster's members and beyond, and even increase growth and development (OECD, 2007).

When these external economies or externalities are self-feeding over time in the territory, they can lead to a process of endogenous growth and development. In addition, this method is substantially diffusive, because its engine is knowledge spillover and, therefore, tends to spread up with its positive effects on other related sectors, due to the proximity of critical technologies, as well as to other territories, first those of geographical contiguity, contributing to their convergence. In this sense, Porter (1991) explains that innovative countries and regions enter into a most dynamic phase of development, spreading up their impact on 'related and support sectors,' once reached maturity in their sectors and branches.

Advantages of location are especially crucial for innovation processes within clusters, as they affect removing barriers or obstacles to innovation, while spatial or geographical proximity favors the reduction of technical and economic uncertainty of innovation processes (Freeman, 1986; Rosenberg, 1982). Besides, the geographic proximity allows a more significant interaction among clusters' stakeholders, which favors the creation of networks and long-term trust relationships, a critical factor in the innovation process (Lundvall, 1992; Von Hippel, 1988). Finally, the innovation process originated in the transference of technology to companies from universities, and technological centers facilitate the proximity among them (Freeman, 1986), while the supply of qualified personnel in its environment catalyzes the innovative process (Rosenberg, 1982).

Moreover, as regards the effects of these economies or location advantages, they can be distinguished two types: those derived from economies of production and operation, which are, above all, improvements in productivity, higher vertical integration, cost reduction, dragging of suppliers, and training within industrial districts. On the other hand, there are the effects derived from innovation economies, which favor innovation and its diffusion, both within the cluster and to related sectors, as well as the impact on technological stakeholders and the attraction of qualified employment.

In any case, the effects of these externalities, and especially those of innovation, are very dependent on the type of innovation processes performed within the cluster, which are in turn highly dependent on the critical sectoral technologies employed.

12.2.5 Cluster Location

Advantages of location in sector clusters are developed from the more significant interaction among close stakeholders in the space of the territory, which often allows creating social capital that favors relationships of trust in the long term, as well as specific attitudes of reciprocity associated with them (Feldstein & Putnam, 2003). These links are fundamental to the innovation process and its diffusion.

In this sense, it is of the most considerable interest the cluster's way of the rollout in the territory, because the region becomes a key factor of integration, organization, and development of its social capital. Thus, the geography of the province, its culture, and the sense of belonging to a community became critical factors for the development of relationships of trust, leading to build up communication networks among stakeholders of the cluster. In this way, advantages of location depend primarily on the agglomeration model and the external economies we have previously discussed, which could become external de-economies, depending on the centrifugal forces of the concentration, which become higher than the centripetal ones, as explains the Krugman–Venables model (Krugman et al., 2000). Similarly, a model of 'cumulative circular causation' developed by Myrdal (1957) and Kaldor (1963) operates in an analogous way to the Krugman–Venables model.

In the dynamics of formation of clusters, specifically in the seeking for advantages or rents of location, spatial phenomena develop and influence on innovation processes within clusters due to the impact of proximity on the diffusion of innovation. Due to this, we can talk about a 'Geometry from Diffusion of Innovation,' and for the aims of our research, we will focus mainly on four types of spatial phenomena provoked by this geometry:

a. *Growth triangles*: natural economic zones of a cross-border nature, formed by bordering areas regarding different endowments of productive factors (Plaza, 1999). Consequently, there are border effects in these areas that allow reducing costs, allocating resources more efficiently and hence to achieve productivity gains. Therefore, it can be said that there is a sub-regional integration that minimizes transaction costs (Coase, 2007[1966]), by improving the 'losers,' which we have already seen that are organized in 'lobbies' or Olsonian groups with low organizational costs (Olson, 1965). As a result, growth triangles are a new approach to regional cooperation and economic growth, as in Asia (Kakazu, 1999).
b. *Trans-border regions*: natural economic zones between areas where a cooperative game is developed, which allows for a better balance solution for all stakeholders (Nash, 1950). They can even give rise to 'Regions-State,' which constitute economies of services with availability of good infrastructures. In this sense, Ohmae (1995) says that depending on their size should allow developing a 'flexible community of interests.'
c. *Nodes and poles of growth*: economic centers of clusters, usually poles or districts, which support the cluster and have a specific centrality over it. Hence, from them, axes or corridors extend along the group.
d. *Axes/corridors*: they start from the nodes of clusters, and they spread advantages of location along them, primarily through processes of technological diffusion.

The emergence of technological clusters that strengthen axes and poles of growth in Mexico complements this geometry. In recent years, Mexico has tried to develop several initiatives to agglomerate technological capacities, but only two (Querétaro and Monterrey) have grown actively in aerospace and nanotechnology clusters due to the application of local policy oriented to technology-based development, and the mobilization of local and global actors in the configuration of technological agglomerations (Luna-Ochoa, Belmont, & Suaste-Gomez, 2016). We will briefly describe the Querétaro Aerospace Valley whose objective is to lead with Brazil the aeronautical market in Latin America.

12.2.6 *The Querétaro Aerospace Valley (Mexico)*

Geometrically, this cluster is a triangle formed by the cities of Querétaro (Mexico), Toulouse (France), Wichita and Seattle (United States), and Montreal (Canada) where more than 3600 specialists work in a network formed by 30 manufactured firms (see Table 12.1).

These corporations are complemented with three MRO (maintenance, repair, and overhaul) firms, three R&D centers, five design and engineer centers, five services firms, three educational institutions, and one R&D network.

One of the keys to explaining the success of this cluster is the public intervention linked to the creation of three-helix ecosystems of entrepreneurship formed by universities, firms, and the State. Examples of this three-helix ecosystem are CIATEQ, CIDESI, and CEDIA. Respecting CIATEQ, as a public research center coordinated by the National Council of Science and Technology (CONACYT), has the role of training high-level human resources related to technological specialties (advanced manufacturing, oil, aeronautic industries, water handling and distribution, automated control systems, mechatronics, and software development), as well as to generate scientific knowledge through applied research. After

Table 12.1 Firms belonging to the Querétaro Aerospace Valley (excluded SMEs)

Company	*Country*	*Company*	*Country*
Bombardier	Canada	Aernnova	Spain
Meggitt	UK	Safran	France
ITR-ITP Group/SENER & Rolls Royce	Spain	General Electric	United States
Dishon Ltd	Israel	Kuo Aerospace	Mexico
LabTA (Aerolabs for Technical Services and Technology)	Mexico	CIDESI (The Center for Engineering and Industrial Development)	Mexico
CIATEQ	Mexico	ELIMCO	Spain
PRETTL	Germany	Navair Technologies	Canada
Drovedani	Italy	Carpenter Aceros Fortuna	United States and Mexico
Hyrsa-American Steel	United States	CRIO	United States
Crowners			
DHI Aero	Mexico		

Source: Authors

38 years of experience, CIATEQ is located in six states (Aguascalientes, Hidalgo, Querétaro, San Luis Potosí, State of Mexico, and Tabasco) and has developed more than 3800 industrial projects and more than 51,000 technological services.

Regarding CIDESI, with sites in Querétaro and the State of Mexico, and the collaboration with the Research and Innovation Technology Park in Nuevo Leon, and the Technology Consortium of Baja California in Tijuana, CIDESI contributes to R&D development in Mexico through strategic alliances with the Autonomous University of Querétaro, the Institute of Astronomy of the National Autonomous University of Mexico, the Aachen University of Applied Societies (Germany), the University of Texas A&M, the University of Lehigh, the University of Sheffield (UK), the University of New Mexico's Anderson School of Management (United States), Tecnalia (Spain), the National Metrology Center, the Japan International Cooperation Agency, the Aeronautical Technologies Center, and Team Technologies (United States). Its relation with industry is outstanding, as CIDESI is a supplier of automotive, auto parts, aerospace, energy, petrochemical, electronics, home appliances, and food industry. This cluster also has a close relationship with the National Aeronautics University in Querétaro (Mexico).

Finally, the CEDIA cluster, formed by Concordia University-CIADI (Concordia Institute of Aerospace Design and Innovation)(Canada), ESTACA École d'Ingénieurs (Canada) and Tecnológico de Monterrey (Mexico) to promote the development of the aerospace cluster in Mexico, training students and technological solutions for the aerospace industry with the objective of integrating research centers and institutions of higher education, with lines of specialty of interest to the aeronautical sector, in order to promote: high level research, and the training of specialists and certified installations for the aeronautical industry.

The challenge facing the aeronautical cluster in Querétaro (Mexico) is to create a sustainable first-order competitive advantage, defined by the combination of R&D and innovation, to make Mexican aeronautical firms to compete on a global scale (Burgos & Johnson, 2016). This process of industrial transformation has mainly been possible due to the positive effects emanated from the combination of the National Development Program 2013–2018, the Pro-Aereo 2012–2020, the State Plan of Development of Querétaro 2016–2021, and the Road Map of the Aerospace Sector for the Region Querétaro 2015.

The good industrial results derived from the aeronautical cluster invite to enlarge the triangle formed by the cities of Querétaro (Mexico), Toulouse (France), Wichita and Seattle (United States), and Montreal (Canada) by inserting the INNOVA model of Madrid Aerospace Cluster (Spain). As a result, one of the most outstanding clusters in the world between Europe and America could be created that could compete with other existing industrial clusters in Asia.

12.2.7 The INNOVA Model of Madrid Aerospace Cluster (Spain)

One example of how innovation runs, grows and is diffused among stakeholders within clusters is the case of the INNOVA model of Madrid Aerospace Cluster, where internal dynamics of clusters, members, and the relationships among them can establish an endogenous development process leading to self-sustained growth, beyond a number of productive resources located in the territory. In this model, knowledge and innovation play a key role, because their diffusion or overflow among cluster members generates externalities that constitute the real motor of endogenous development within it (Grossman & Helpman, 1991). Furthermore, innovation in the cluster endogenizes the progress and, as a consequence, maintains a constant growth to exceed the decreasing returns characteristic of the sizable factorial increase, and thus not entering into a steady-state situation.

The INNOVA model is based on the relationships between the five critical factors that develop the aerospace cluster, as follows:

a. Competitiveness of the territory as a vital element in the formation of the industrial cluster.
b. Innovation and social capital as cohesive drivers of the cluster.
c. Development of a regional innovation system in the territory.
d. The waterfall diffusion process of innovation in the cluster.
e. The structure of the INNOVA model of Madrid Aerospace Cluster deployed in the physical (geographical) territory.

We will analyze this cluster by using a factorial analysis described in next paragraphs.

12.3 Method

For this research work, we have sent a questionnaire to cluster companies and conducted interviews with relevant stakeholders. The design of the research is qualitative and quantitative, as we have combined open-talk interviews done in the Madrid Aeronautical Cluster, the public administration, universities, and research centers to identify some critical issues. A questionnaire of 23 closed questions sent to all 53 members of the Madrid Aerospace Cluster, the Aeronautical Sectoral Association (ATECMA) and the Space Sectoral Association (ProEspacio) complemented this qualitative data. We have received answers from 36 companies, representing the entire industrial value chain, including some service companies to the industry. From the analysis of data obtained in the questionnaires and interviews, we were able to contrast the importance of locational factors for the regional development, and it has been possible to define and validate the INNOVA Madrid Aerospace Cluster model.

With data obtained from the questionnaires, we identified which are the business variables related to R&D and innovation processes carried out by the INNOVA Madrid Aerospace sector. To further deep into the relationships among variables of the sample, we performed a factor analysis, which has allowed us to confirm more healthy causality relationships not directly observable, as the size of companies and their intellectual capital, or their size and the percentage of equity belonging to other companies in the sector. In this way, we contrasted the INNOVA model, because the innovation pattern of corporations closely relates to their intellectual capital.

Next, we perform a cluster analysis based on the data collected in the questionnaire to locate the companies that are part of our research, to form groups of them within the sample. This fact helps us to distinguish among types of companies within the sector, as well as to contrast the industrial structure organized into groups, according to their position in the value chain. Thanks to this analysis, we will identify, with a relatively small number of factors, the extensive and complex relationships that can exist between a set of observed variables. In this respect, we use two cluster analysis procedures, available on SPSS v.19: hierarchical analysis and K-means analysis. The hierarchical analysis is ideal to determine an optimum number of clusters or groups in the data and their contents, while the K-means analysis allows processing an unlimited number of cases, but using only one grouping method, and requires that the number of clusters to be obtained had been previously proposed.

12.3.1 *Research Results*

We apply the factorial analysis because it is a statistical technique used to identify, with a relatively small number of factors, the extensive and complex relationships existing between a set of observed variables. That is, the factorial technique tries to find a group of elements, with various but not directly observable dimensions, that may exist among a large number of visible variables (Pardo & Ruiz, 2002).

The factorial analysis consists of four necessary steps: (1) to calculate the matrix able to collect the joint variability of all the selected variables, (2) to extract the optimum number of factors, (3) to rotate the solution to facilitate its interpretation, and (4) to estimate the factor scores in the new rotated dimensions. For the analysis, we have selected five variables, and in case there was a correlation between them, they would allow us to confirm the proposed model. Explicitly, the selected variables and their basic statistics are shown in Table 12.2.

To carry out correlation analysis among variables we used the Pearson correlation coefficient between each pair of variables. In addition, to find common factors among observed variables we must see degrees of multicollinearity between them, that is to say, statistical correlation levels that were significant. Concerning this latter, we show in Table 12.3 the correlation coefficients for the variables considered, as well as their degree of significance. We can see that some variables correlate with others with values above 0.500, that is, with a significant level of correlation, which also includes the unilateral critical level (Unilateral Sig). Furthermore, a critical level less than 0.05 points out that correlation between the corresponding pair of variables is significantly different from zero, so that conditions to continue the analysis are given.

Table 12.2 Descriptive statistics of selected variables from the sample

Variable	*Statistical mean*	*Standard deviation*	*Sample size*
Fact_Total	10.99	6.93	36
Empleo	335.89	240.55	36
Cap_Sector	25.28%	24.98%	36
Emple_cualif	91	100	36
FTE_Idi	39	29	36

Source: Authors

Table 12.3 Matrix of correlations

		Fact_Total	*Empleo*	*Cap_Sector*	*Emple_cualif*	*FTE_Idi*
Correlation	Fact_Total	1.000	0.898	0.268	0.674	0.750
	Empleo	0.898	1.000	0.178	0.573	0.765
	Cap_Sector	0.268	0.178	1.000	−0.031	0.031
	Emple_cualif	0.674	0.573	−0.031	1.000	0.648
	FTE_Idi	0.750	0.765	0.031	0.648	1.000
Sig. (Unilateral)	Fact_Total		0.000	0.057	0.000	0.000
	Empleo	0.000		0.150	0.000	0.000
	Cap_Sector	0.057	0.150		0.428	0.428
	Emple_cualif	0.000	0.000	0.428		0.000
	FTE_Idi	0.000	0.000	0.428	0.000	

Source: Authors

In addition, we performed the so-called Kaiser–Meyer–Olkin (KMO) correlation test, which is an index that allows comparing the magnitude of the correlation coefficients observed with the importance of the partial correlation coefficients, and which has the following expression:

$$KMO = \frac{\sum r_{ij}^2}{\sum r_{ij}^2 + \sum r_{ij,m}^2}$$

Where r_{ij} represents the simple correlation coefficient between variables i and j, and $r_{ij}.m$ represents the partial correlation between variables i and j, after having eliminated the effect of the remaining m variables included in the analysis. Since the partial correlation between two variables should be small when the factorial model is adequate, the denominator should increase little if the data correspond to a factorial structure, in which case KMO takes a value close to 1.

In this context, partial correlation coefficient expresses the degree of relationship between two variables, after eliminating the effect of the remaining variables included in the analysis. Thus, when the variables included in the analysis share a significant amount of information, due to the presence of common factors, the partial correlation between any pair of variables should be reduced. Thus, small values of the index express that there is a sizable partial correlation between two variables, but not shared with the remaining variables, being this a lousy symptom regarding the suitability of the analysis and indicating that it would not be a good idea to proceed further with factor analysis. The value taken from the KMO index in this research is 0.719, which in this way can be interpreted as correct. In addition, Bartlett test confirms that data collected does not generate the same matrix with a 0.000 significance so that we can go on with the analysis (see Table 12.4).

Then, once we have checked linearity between selected variables, we can perform the extraction of factors that justify observed correlation. To do so, we apply the principal component analysis, which forms linear combinations of observed variables and allows us to see which components of taken variables explain a more considerable amount of their total variance, that is, lets explain better dispersion with regard to the group mean, as do Buesa et al. (2003). So as, this algorithm is successively grouping variables observed in groups two by two with the correlation between them (and without correlation with other groups), and thus can explain the amount of total variance due to each factor, regarding the full standard variance to explain.

Table 12.5 contains commonalities initially allocated to the variables, as well as commonalities reproduced by the factor solution (extraction). The

Table 12.4 KMO and Bartlett tests

Sample Adequacy Kaiser–Meyer–Olkin		*0.719*
Bartlett Sphericity Test	Chi-square proxy	114.799
	gl	10
	Sig.	0.000

Source: Authors

Table 12.5 Commonalities

	Initial	*Extraction*
Fact_Total	1.000	0.914
Empleo	1.000	0.851
Cap_Sector	1.000	0.965
Emple_cualif	1.000	0.703
FTE_Idi	1.000	0.806

Source: Authors

commonality of a variable is the proportion of its variance that can be explained by the obtained factor model. The total percentage of the variance that is described by each variable is equal to the sum of the dimensions of the variance explained by each factor. By studying the commonalities of the extraction, it is possible to assess which variables are the worst described by the model. In this sense, in our model variable *Emple_cualif* is the worst defined, since the model is only able to reproduce 70.3% of its original variability.

From the above analysis, we obtained two eigenvalues greater than 1, so we observed that the procedure had extracted two factors that manage to explain 84.77% of the variance of the original data. Moreover, the first element reaches a value of 63.77%, equivalent to 3.188% of the standardized total value.

The extraction of rotated values simplifies and re-balances the value ratio of each of the factors, being now 62.83% the explanation of the variance made by the first factor, and 21.95% that of the second. In sum, the five primary observed variables could be reduced to two elements, which explain 84.77% of the total variance without losing significant information.

Once we know that two factors are enough to explain reasonably well the relationship between observed variables, we must next estimate coefficients that relate elements to variables. For this aim, we use the principal components method, which allows us to determine coefficients among original variables (or saturations) and each of the factors. Moreover, it is necessary to reduce the number of variables in the first factor that has a high factor load (saturation), by applying an orthogonal rotation method known as Varimax.

The matrix of rotated components allows us to assign the five variables to each of the two factors, so we can then 'name' the elements based on the variables contained inside. This is the phase of analysis most conditioned by the theoretical scheme of the statistical model exposed, given that from the interpretation made of the factors with the variables contained, it is possible to extract conclusions from the analytical model and determine whether confirm or not some of the leading starting hypotheses.

The first factor is called factor dimension and intellectual capital, which contains the following variables: turnover of industrial activity in the cluster, number of employees, number of qualified employees, and number of employees with Full-Time Equivalents (FTEs) dedicated to R&D and innovation. The reason for calling it so is due to the relationship between variables such as the size of the companies, measured either by sales or by the number of employees, and intellectual capital (as a proxy of knowledge) in them.

The second factor, composed of the variable 'Percentage of Capital belonging to other Companies in the Sector,' is called capital factor, given that it provides us with information about cross-participation between companies in the aerospace sector in the region of Madrid.

At this point, we can interpret the factor analysis' statistical model to reach two basic conclusions:

a. The dimension factor and intellectual capital show that, for the region of Madrid's aerospace industrial cluster, there is a high correlation between the size of the companies and the amount of existing knowledge. In other words, a higher level of production gives a higher degree of technological expertise in corporations, so a high level of technical knowledge can be reached by the sector (Dosi, Giuri, & Tomasi, 2005), considering that 72% of the companies surveyed have a turnover higher than 7 million euros.
b. The capital factor shows the relatively high integration between the structures of capital in the sector. As we have seen, wealth tends to be very concentrated and, in turn, with a high share of cross-interests within the group of companies on TIER 1 and TIER 2.

The combination of knowledge and capital concentration reinforces the Spanish INNOVA model of Madrid Aerospace Cluster and incentives to search for new business opportunities across the Atlantic and in other regions of the world.

12.4 Conclusions

From the objective of this work, results are fivefold. First, data results show that it will be favorable to the Querétaro Aerospace Valley (Mexico) to be inserted into the INNOVA model of Madrid Aerospace Cluster (Spain) to create an axis of growth formed in America by Mexico (Querétaro and Guadalajara), the United States (Wichita and Seattle) and Canada (Montreal), and Europe by France (Toulouse) and Spain (Madrid). As a result, economic growth and social welfare will increase, so social imbalances will tend to diminish. The creation of this axis of development will lead to a useful improvement of the populations living in these territories, which will benefit firms located in other regions of Mexico and will allow competing against other aeronautical clusters existing in other areas in the world.

Second, by having Spain and Mexico the same language facilitates the exchange of experiences, labor integration, the creation of work teams, and collaboration between highly specialized human resources. This more significant team integration is due to the absence of cultural and linguistic barriers. As a result, the cluster strengthens and leads to a long-term relationship.

Third, aerospace clusters show how a broader private-public sector collaboration can enhance the standard of living of the populations living in the augmented cluster (Mexico–United States–Canada–France–Spain) due to the crowding-in effect. In fact, when the private company has the support of the level to develop an industry, a multiplier effect in other sectors is formed to impulse the economy for generating growth. As a result, the strategic and policy implications of our work will link with public initiatives focused on creating economic welfare to reduce human poverty in emerging economies, and particularly in Mexico.

Fourth, the combination of different aerospace-based clusters strengthens the impact of innovation-related processes to establish an endogenous development process leading to self-sustained growth, beyond the localized productive resources in the territory, where knowledge and innovation play a crucial role. As a result, and rooted in diffusion or overflow processes among the cluster, stakeholders generate externalities that constitute the real engine of endogenous development.

Fifth, this new reinforced aerospace-based cluster will:

a. Reinforce technological-rooted knowledge in specialized firms
b. Strengthen the importance of engineering activities in civil aviation markets
c. Increase technological understanding of materials, composites, and aerostructures
d. Augment the cross-participation in the capital of cluster companies, as 30% of them have majority interests in other companies
e. Upsurge the technological drag effect of suppliers, and
f. Lead to a niche technical leadership by some cluster SMEs due to complementary knowledge on some critical sectoral technologies.

Finally, the combination of both industrial clusters will strengthen the aerospace clusters located in Europe and America to achieve sustainable growth rates that can contribute, in the Mexican case, to reduce poverty and to build an adequate infrastructure and transportation system. As a result, firms working in the cluster will have better access to strategic inputs, capital, inputs, and knowledge that will attract professionals and corporations with the enormous potential to take advantage of them by transforming the countries belonging to the cluster to benefit future generations.

References

Baines, T. S., Lightfoot, H. W., Benedettini, O., & Kay, J. M. (2009). The Servitization of Manufacturing. A Review of Literature and Reflection on Future Challenges. *Journal of Manufacturing Technology Management, 20*(5), 547–567.

Becattini, G. (1979). Dal settore industriale al distretto industriale: alcune considerazione sull'unita di indagine dalla economia industrial [From the Industrial Sector to the Industrial District: Some Considerations on the Research Uniqueness of the Industrial Economy, original in Italian]. *Rivista di Economia e Politica Industriale, 1*, 7–21.

Becattini, G. (1989). *Modelli locali di sviluppo* [Local Models of Growth, original in Italian]. Il Mulino, Bologna, Italy.

Becattini, G. (2017). The Marshallian Industrial District as a Socio-Economic Notion. *Revue d'Economie Industrielle, 157*(1), 13–32.

Behera, S. R. (2017). Regional Foreign Direct Investment and Technology Spillovers: Evidence Across Different Clusters in India. *Economics of Innovation and New Technology, 26*(7), 596–620.

Bellantuono, N., Carbonara, N., & Pontrandolfo, P. (2017). The Organization of Eco-industrial Parks and Their Sustainable Practices. *Journal of Cleaner Production, 161*, 362–375.

Bergek, A., Jacobsson, S., Carlsson, B., Lindmarki, S., & Rickne, A. (2005). *Analysing the Dynamics and Functionality of Sectoral Innovation Systems–A Manual.* In 10 Year Anniversary DRUID Summer Conference, Copenhagen, June 27–29.

Bernard, C. I. (1938). *The Functions of the Executive.* Boston, MA: Harvard University Press.

Bowen, H. R. (1953). *Social Responsibilities for the Businessman.* New York: Harper.

Buesa, M., Baumert, Th., Heijs, J., & Martínez Pellitero, M. (2003). Los factores determinantes de la innovación: un análisis econométrico sobre las regiones españolas. Economía Industrial. En prensa.

Burgos, R. G., & Johnson, J. (2016). Why Querétaro? The Development of an Aeronautical Manufacturing Cluster in Central Mexico, Thunderbird International Business Review, in press.

Carlsson, B., & Stankiewicz, R. (1991). On Nature, Function, and Composition of Technological Systems. *Journal of Evolutionary Economics, 1*, 93–118.

Carlsson, B., Jacobsson, S., Holmén, M., & Rickne, A. (2002). Innovation Systems: Analytical and Methodological Issues. *Research Policy, 31*, 233–245.

Castells, M., & Hall, P. (1994). *Las tecnópolis del mundo* [Technopolis in the World, original in Spanish]. Madrid, Spain: Alianza Editorial.

Coase, R. H. (2007[1966]). *La empresa, el mercado y la ley* [The Firm, The Market, and The Law, original in English]. Barcelona, Spain: Alianza Editorial.

Cooke, P., & Morgan, K. (1994). The Creative Milieu: A Regional Perspective on Innovation. In M. Dodgson & R. Rothwell (Eds.), *The Handbook of Industrial Innovation* (pp. 25–32). Aldershot, UK: Edward Elgar.

Cossentino, F., Pyke, F., & Sengenberger, W. (1996). Local and Regional Response to Global Pressure: The Case of Italy and Its Industrial Districts, World Trade Organization, Genève, Switzerland.

COTEC Foundation. (2017). COTEC 2017 Report on the Spanish Innovation System [from Spanish, Informe COTEC 2017 Sobre el Sistema Español de Innovación], Madrid, Spain.

Crociata, A., Furia, D., & Agovino, M. (2018). The Cognitive Determinants of Social Capital. Does Culture Matter? *Advances in Intelligent Systems and Computing, 618*, 105–113.

Davis, K. (1960). Can Business Afford to Ignore Social Responsibilities? *California Management Review, 11*(3), 70–76.

Davis, K. (1967). Understanding the Social Responsibility Puzzle: What Does the Businessman Owe to Society? *Business Horizons, 10*(4), 45–50.

Dini, M. E. (1996). Políticas públicas para el desarrollo de redes de empresas. La experiencia chilena [Public Policies to Develop Business Networks. The Chilean Experience, original in Spanish]. *Revista Latinoamericana de Estudios del Trabajo, 2*(3), 131–158.

Dosi, G., Giuri, P., & Tomasi, C. (2005). The European Aerospace Industry. Industrial Structure and Knowledge Spillovers. Presented at Cotec Europe Conference, Lecce (Italy).

Edquist, C. (2005). Systems of Innovation: Perspectives and Challenges. In J. Fagerberg, D. C. Mowery, & R. R. Nelson (Eds.), *The Oxford Handbook of Innovation* (pp. 181–208). New York: Oxford University Press.

Etzkowitz, H. (2014). Making a Humanities Town: Knowledge-infused Clusters, Civic Entrepreneurship and Civil Society in Local Innovation Systems, Triple Helix (Springer Open Journal), pp. 1–12.

European Cluster Observatory. (2009). EU Cluster Mapping and Strengthening Clusters in Europe. Europe INNOVA Paper No. 12. http://publications.europa.eu/resource/cellar/6f14c45f-7d6a-49c7-9bbf-785b313657d4.0001.03/DOC_2

European Commission. (2008). Europe INNOVA: European Innovation Platforms. http://ec.europa.eu/growth/content/europe-innova-european-innovation-platforms_en

Feldstein, L., & Putnam, R. (2003). *Better Together: Restoring the American Community*. New York: Simon and Schuster.

Freeman, C. (1986). *The Economics of Industrial Innovation* (second ed.). Cambridge, MA: MIT Press.

García-López, M. A., & Muñiz, I. (2010). El impacto espacial de las economías de aglomeración y su efecto sobre la estructura espacial del empleo. El caso de la industria en Barcelona (1896–1996), España [The Impact of Space Agglomeration Economies and its Effects on the Social Structure of Employment. The Case of the Industry in Barcelona (1896–1996), Spain, original in Spanish]. *Revista de Economía Aplicada, 18*(52), 91–119.

Gray, M. (1996). Big Firms, Long Arms, Wide Shoulders: 'Hub-and-Spoke' Industrial District in the Seattle Region. *Regional Studies, 30*(7), 651–666.

Grossman, G. M., & Helpman, E. (1991). Quality Ladders in the Theory of Growth. *Review of Economic Studies, 58*(1), 43–61.

Guimón, J., & Paraskevopoulou, E. (2017). Factors Shaping the International Knowledge Connectivity of Industrial Clusters: A Comparative Study of Two Latin American Cases, Entrepreneurship, and Regional Development, in press.

Johanson, J., & Vahlne, J.-E. (1977). The Internationalization Process of the Firm—A Model of Knowledge Development and Increasing Foreign Market Commitments. *Journal of International Business Studies, 8*(1), 23–32.

Kakazu, H. (1999). Growth Triangles in Asia. A New Approach to Regional Cooperation, Working Paper, 9, Politics & International Relations Series, International University of Japan, Niigata, Japan.

Kaldor, N. (1963). Capital Accumulation and Economic Growth. In F. A. Lutz, C. H. Douglas (Eds.). Proceedings of a Conference held by the International Economics Association, Macmillan, London, UK.

Kreps, T. J. (1940). Measurement of the Social Performance of Business. *The Annals of the American Academy of Political and Social Science, 343*(1), 20–31.

Krugman, P. (1991a). *Geografía y Comercio* [Geography and Commerce, original in English]. Barcelona, Spain: Antoni Bosch Editor.

Krugman, P. (1991b). Increasing Returns and Economic Geography. *Journal of Political Economy, 99*, 483–499.

Krugman, P., Fujita, M., & Venables, A. J. (2000). Economía Espacial, Ariel, Barcelona, Spain.

Luna-Ochoa, S. M. A., Belmont, E., & Suaste-Gomez, E. (2016). A Profile of Mexico's Technological Agglomerations: The Case of the Aerospace and Nanotechnology Industry in Querétaro and Monterrey. *Technology in Society, 46*, 120–125.

Lundvall, B. (1992). *National Systems of Innovation: Toward a Theory of Innovation and Interactive Learning*. London, UK: Anthem Press.

Markusen, J. R. (2002). *Multinational Firms and the Theory of International Trade*. Cambridge, MA: MIT Press.

Martí Fluxá, R. (2005). Plan of the Aerospace Sector of the Community of Madrid [from Spanish, Plan del Sector Aeroespacial de la Comunidad de Madrid]. Unpublished manuscript, Madrid, Spain.
Milian, L. (2015). *Responsabilidad Social Corporativa. Origen y Evolución en el Entorno Empresarial Europeo y Español* [CSR. Origin and Evolution in the Spanish and European Business Environment, original in Spanish], Madrid, Spain: Pontifical University of Comillas.
Myrdal, G. (1957). *Economic Theory and Under-Developed Regions.* London, UK: Gerald Duckworth and Co..
Nash, J. (1950). Equilibrium Points in n-person Games. *Proceedings of the National Academy of Sciences, 36*(1), 48–49.
Newbert, S., & Craig, J. B. (2017). Moving Beyond Socioemotional Wealth: Toward a Normative Theory of Decision Making in Family Business. *Family Business Review, 30*(4), 339–346.
OECD. (2007). OECD Reviews of Innovation Policy OECD Reviews of Innovation Policy. Competitive Regional Clusters.
Ohmae, K. (1995). *The End of the Nation State. The Rise of Regional Economies.* New York: The Free Press.
Olson, M. (1965). *The Logic of Collective Action: Public Goods and the Theory of Groups.* Boston, MA: Harvard University Press.
Pardo, A., & Ruiz, M. A. (2002). *SPSS 11. Guía para el análisis de datos.* Madrid, Spain: McGraw-Hill.
Perroux, F. (1955). Notes sur la notion de pôle de croissance [Some Notes on the Concept of Growth Pole, original in French]. *Economie Appliquée, 7*, 307–320.
Perroux, F., 1960. *Economie et Societé. Contrainte, Echange* [Economy and Society. Constraint and Change, original in French]. Paris, France: Dons.
Plaza, S. (1999). Grandes ciudades y medioambiente [Big Cities and Environment, original in Spanish]. In M. A. Díaz Mier & M. A. Galindo (Eds.), *Economía y Medioambiente.* Madrid, Spain: Biblioteca Nueva.
Porter, M. (1991). *The Competitive Advantage of Nations.* New York, USA: The Free Press.
Porter, M. (2003). The Economic Performance of Regions. *Regional Studies, 37*, 549–578.
Prive, T. (2012). Top 10 Qualities That Make a Great Leader, Forbes, Dec 19.
Pu, H., & Zheng, Y. (2015). The FDI of Small-and-Middle-Sized Enterprises: A Literature Review. *Technology and Investment, 6*(1), 1–7.
Rabelloti, R. (1997). *External Economies and Cooperation in Industrial Districts. A Comparison of Italy and Mexico.* London, UK: Macmillan.
Rabelloti, R. (2003). The Effect of Globalization on Industrial Districts in Italy: The Case of Brenta. In H. Schmitz (Ed.), *Globalised Localities.* London, UK: Edward Elgar.

Ramírez, D. N. (2016). *The Social Mortgage [original en español, La Hipoteca Social]*. Mexico: McGraw-Hill.

Ranga, M., & Etzkowitz, H. (2013). Triple Helix Systems: An Analytical Framework for Innovation Policy and Practice in the Knowledge Society. *Industry and Higher Education, 27*(4), 237–262.

Rosenberg, N. (1982). *Inside the Black Box*. New York: Cambridge University Press.

Saiz-Álvarez, J. M. (2017). Entrepreneurial ICT-based Skills and Leadership for Business Ethics in Higher Education. In S. Mukerji & P. Tripathi (Eds.), *Handbook of Research on Technology-Centric Strategies in Higher Education Administration* (pp. 373–392). Hershey, PA: Information Science Reference (IGI Global).

Saiz-Alvarez, J. M., & García-Ochoa, M. (2008). Externalización de Servicios y Alianzas Estratégicas en la Nueva Economía del Conocimiento [Outsourcing and Strategic Alliances in the New Knowledge Economy, original in Spanish]. *Economía Industrial, 370*, 135–141.

Saxenian, A. L. (1994). *Regional Advantage: Culture and Competition in Silicon Valley and Route 128*. Cambridge, MA: Harvard University Press.

Schwartz, M. S., & Carroll, A. B. (2003). Corporate Social Responsibility: A Three-Domain Approach. *Business Ethics Quarterly, 13*(4), 503–530.

Sharma, M. G., & Singh, K. N. (2017). Servitization, Coopetition, and Sustainability: An Operations Perspective in Aviation Industry. *Vikalpa, 42*(3), 145–152.

Sölvell, O. (2008). *Clusters: Balancing Evolutionary and Constructive Forces*. Sweden, Ivory Tower.

Vandermerwe, S., & Rada, J. (1988). Servitization of Business: Adding Value by Adding Services. *European Management Journal, 6*(4), 314–324.

Vázquez-Barquero, A. (1999). *Desarrollo, Redes e Innovación. Lecciones sobre Desarrollo Endógeno* [Growth, Networks, and Innovation. Some Lessons on Endogenous Growth, original in Spanish. Madrid, Spain: Ediciones Pirámide.

Von Hippel, E. (1988). *The Sources of Innovation*. Oxford, UK: Oxford University Press.

Wells, L. T. (1983). *Third World Multinationals: The Rise of Foreign Investments from Developing Countries*. Cambridge, MA: The MIT Press.

CHAPTER 13

Strategies for Identification of Interdisciplinary Research Subjects: Approach, Tools, and Opportunities for Sustainable Development of Agricultural Economies

Wilmar Osorio Viana, Paola Andrea Calderón Cuartas, and Javier Mauricio Naranjo Vasco

13.1 Introduction

Investment in science, technology, and innovation (ST&I) is a fundamental issue for the nations around the world because it is a key factor for economic growth and social development: it constitutes one of the main promoters of productivity and competitiveness of nations. There is a lot of information and worldwide statistics available related to scientific production, productivity, quality, and social impact of scientific research,

---

W. Osorio Viana (✉) • P. A. Calderón Cuartas • J. M. Naranjo Vasco
Research Group in Technological and Environmental Developments, Department of Environmental Engineering, Faculty of Engineering and Architecture, Universidad Católica de Manizales, Manizales, Colombia
e-mail: wosorio@ucm.edu.co; pcalderon@ucm.edu.co; jnaranjo@ucm.edu.co

C. I. Pardo Martínez et al. (eds.), *Analysis of Science, Technology, and Innovation in Emerging Economies*,
https://doi.org/10.1007/978-3-030-13578-2_13

technological development, and innovation (R&D&I) efforts by country Ahmadpoor & Jones 2017; Cornell University, INSEAD & WIPO 2017). In such a landscape of interesting and often amazing figures, there is a constant fact since the beginning of the twentieth century until the latest reports in 2017: a poorer performance of Latin America and the Caribbean (LA&C) countries compared to United States of America and the European Union. Scientific discoveries, technological inventions, and industry innovations, as measured by the number of top journal articles, patents, industrial revenues, and human well-being and quality of life indexes, are much less than that in the developed countries. Reasons to explain this situation are often found in public and private expenditure in R&D&I efforts, as a positive correlation between gross domestic product (GDP) and R&D&I success is yet to be proven (Cornell University et al. 2017). Even a positive correlation between GDP and Human Development Index (HDI) has also been proven (United Nations Development Programme, 2016). See Table 13.1.

The Global Innovation Index (GII) is an annual ranking of countries by their capacity for, and success in, innovation. Cornell University, European Institute of Business Administration (INSEAD), and the World Intellectual Property Organization, in partnership with other organizations and institutions publish it. Table 13.1 shows some countries in the GII 2017 ranking. In this report, GII provides detailed metrics about the innovation performance of 127 countries and economies. Its 81 indicators explore a broad vision of innovation, including political environment, education, infrastructure, and business sophistication. The first 20 positions correspond to countries with high incomes and that belong to the European community and North America. Chile is the best-ranked country in Latin America and the Caribbean in position 46th, and Colombia is in the 65th position, being the 5th country in the region after Costa Rica, Mexico, and Panama.

Lagging in economic development is related to weaker technological capacity, whereas strong economic development indicates increasing technological capacities. Well-known success stories are those of Japan and the "Four Asian Tigers" (economies of Hong Kong, Singapore, South Korea, and Taiwan), who managed to very rapidly transform from technologically lagging, low-cost manufacturing bases into new global hubs of innovation. There is now increasing attention to whether or not emerging economies may replicate such a transformation in the domain of clean-tech innovation and environmental sustainability (Gosens, Lu, & Coenen, 2015).

There is an apparent positive relationship between the total factor of productivity of a country (TFP) and the Global Innovation Index. As stated previously, there is an apparent positive relationship between innovation and

Table 13.1 Some countries in the Cornell University et al. 2017 ranking (adapted)

Country	*GII rank*	*Income*	*GDP*[a]	*Region*	*Human Development Index (HDI)*[b]
Switzerland	1	HI	708.8	EUR	0.939
Sweden	2	HI	594.98	EUR	0.913
Netherlands	3	HI	891.04	EUR	0.924
United States of America	4	HI	20,200	NAC	0.92
United Kingdom	5	HI	2660	EUR	0.91
Denmark	6	HI	351.56	EUR	0.925
Singapore	7	HI	316.87	SEAO	0.925
Finland	8	HI	271.82	EUR	0.895
Germany	9	HI	3930	EUR	0.926
Ireland	10	HI	357.17	EUR	0.923
Korea, Rep	11	HI	1600	SEAO	0.901
Luxembourg	12	HI	68.58	EUR	0.898
Iceland	13	HI	28.79	EUR	0.921
Japan	14	HI	5060	SEAO	0.903
France	15	HI	2770	EUR	0.897
Hong Kong (China)	16	HI	344.78	SEAO	0.917
Israel	17	HI	361.61	NAWA	0.899
Canada	18	HI	1760	NAC	0.92
Norway	19	HI	406.46	EUR	0.949
Austria	20	HI	442.42	EUR	0.893
Chile	46	HI	265.22	LCN	0.847
Costa Rica	53	UM	61.06	LCN	0.776
Mexico	58	UM	1250	LCN	0.762
Panama	63	UM	63.68	LCN	0.788
Colombia	65	UM	322.49	LCN	0.727
Uruguay	67	HI	65.82	LCN	0.765
Brazil	69	UM	2200	LCN	0.754
Perú	70	UM	221.71	LCN	0.74
Argentina	76	UM	639.22	LCN	0.827
Jamaica	84	UM	14.96	LCN	0.73
Paraguay	85	UM	30.56	LCN	0.693
Trinidad y Tobago	91	HI	20.8	LCN	0.78
Ecuador	92	UM	99.69	LCN	0.739

From Cornell University et al. 2017 report. *UM* upper-middle income, *HI* high income. Regions are based on the United Nations Classification: *EUR* Europe, *NAC* Northern America, *LCN* Latin America and the Caribbean, *SEAO* South East Asia, East Asia, and Oceania, *NAWA* Northern Africa and Western Asia, *SSF* Sub-Saharan Africa

[a]GDP: Gross Domestic Product in Billions of U.S. Dollars (Source: International Monetary Fund, 2017)
[b]Human Development Data 2016 (Source: United Nations Development program, Human Development Reports)

gross domestic product (GDP) per capita. Greater innovation capabilities allow economies to add value and increase exports of medium- and high-technology products (Cornell University et al. 2017). Economies with greater capacities in science, technology, and innovation are considered knowledge-based economies (Consejo Nacional de Política Económica y Social-CONPES, Dirección Nacional de Planeación de Colombia-DNP, 2015), and the economies included in this category tend to have a more stable and sustainable growth, less environmental impacts, and more efficient use of natural resources. However, many of the countries with emerging economies have not yet recognized the importance of STI with a very low investment in these topics: up to 2.38% in average of the GDP. In countries such as South Korea and Israel, the investment in STI is up to 4% of GDP (note that some emerging economies have a GDP about the same order of magnitude of some developed countries, see Table 13.1). In the case of Colombia, it was found that R&D&I investment is lower compared to international and even regional referents. In 2016, this investment was around 0.27% of the GDP. In order to increase the investment at the regional level, in 2011 the Science, Technology and Innovation Fund was created within the framework of the General Royalties System (earnings from the exploitation of natural resources) assigning up to 10% of these resources for R&D&I (CONPES, DNP, 2015). Colombian is an agricultural economy, based on land farming, minerals-fossil fuels exploitation, and food production, with significant indexes of poverty, social and gender inequality, violence, corruption, and environmental contamination (United Nations Development Programme, 2016). The problems of economic, social, cultural, and environmental development in regions of agricultural tradition as Latin America and the Caribbean require a research approach that takes advantage of the latest advances in science, technology, and innovation considering the particularities of the context. Therefore, research groups require applying analytical and systemic approaches to knowledge management in order to make intelligent decisions in relation to the selection of lines of research, projects, resources, and strategies of collaborative work.

Through this chapter, we are going to explore a proposed strategy to identify the most pertinent interdisciplinary research themes using the background of a research group as well as visualizing links between main actors in a science, technology, and innovation system: higher education institutions, productive sector, and government. We analyze a research group in the field of engineering from the Catholic University of Manizales (Colombia) as a case study. The proposed approach allows the application of knowledge management indicators to guide the development of research efforts. It provides elements to understand the particularities of current research systems in developing countries and ways to enhance them to generate high-impact knowledge products in order to contribute to the transformation of our societies.

13.2 Literature Review

Technological innovation systems (TIS) are defined as "dynamic networks of agents interacting under a particular institutional infrastructure and involved in the generation, diffusion, and utilization of technology" (Bergek et al. 2015). The components of an innovation system are the actors, networks, and institutions contributing to the overall function of developing, diffusing, and utilizing new products (goods and services) and processes (Bergek et al. 2008, 2015).

Colombian model for science, technology, and innovation is based on the TIS model and it is focused on three main components (Fig. 13.1): (1) knowledge generators (includes, among others, higher education

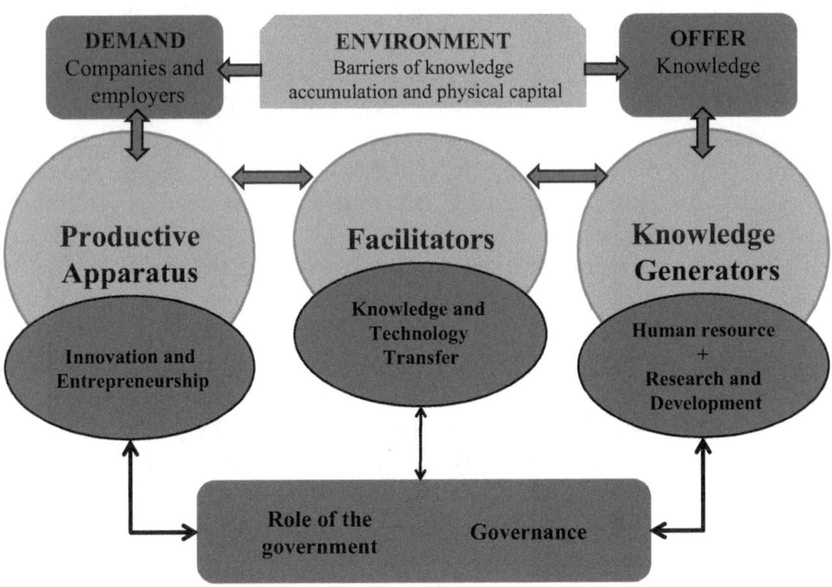

Fig. 13.1 Colombian model for science, technology, and innovation (Adapted from CONPES, DNP, 2015)

institutions (HEI) and R&D centers, as well as companies that generate knowledge); (2) productive apparatus (refers to companies, productive units and, entrepreneurs, the responsible agents for the innovation process, and the increasing of the productivity); (3) facilitators (refers to the dynamics and actors that allow overcoming the barriers of accumulation and allocation of physical capital and knowledge) (CONPES, DNP, 2015).

Despite of Colombia has a defined model for science, technology, and innovation, the main problem is that the country and its regions have not been able to gain economic and social growth as a consequence of the following specific problems: (1) low generation of high-impact knowledge; (2) insufficient conditions for cooperation between the public and private sectors in the identification, demand, and appropriation of knowledge and technology; (3) low innovative and entrepreneurial activities in the productive apparatus (CONPES, DNP, 2015).

In a general perspective, the low generation of high-impact knowledge is closely related to the problem of identifying new strategic research ideas. The later can be approached from different theories including complex systems (Zeng et al., 2017; Fortunato et al., 2018), interdisciplinarity (Frodeman, 2017), big data science (Pentland, 2012), meta-analysis (Gurevitch et al., 2018), and cognitive maps (Gray et al., 2013). If innovation is considered as a guiding principle in knowledge generation and management, the theory of innovation systems will allow the interpretation of both external and internal factors, as well as the identification of actors and their relationships in an innovation system, as proposed by Bergek et al. (2015). In this case, research teams are seen as actors in a given innovation system.

The theory of mental models is used as a simple and practical tool to define and visualize the relationships between the constitutive elements of the analysis and to approach the synthesis of new representations of reality (Henly-Shepard et al., 2015). Consequently, this kind of model was used as a guide for "a model for action" in relation to competitiveness in developing countries as follows: (1) as an explanation of the particularities of several case studies in the 1990–1996 decades (Fairbanks & Lindsay, 1997), and (2) as a proposed tool to induce the change of "mentality" of a nation (Fairbanks & Lindsay, 1997).

There is a great diversity of studies related to scientific productivity and other measures of performance in research activities (such as the 2017 Journal Citation Reports from Web of Science). In these studies, the method is commonly supported by the use of statistical tools for the analysis of relationships and correlations between indicators or indexes (such as the so-called impact factor of scientific publications or the h-index of a researcher) with other factors related to research activities. Such studies usually make use of databases such as Scopus, Web of Science, Scimago Journal Rankings, or even social networks as LinkedIn or Mendeley to process big data about groups, researchers, and research subjects, without considering the internal peculiarities of each of these groups. There are few studies focused on the structural and historical analysis of the research teams in order to propose adaptive knowledge management strategies considering the specific cultural and geographical context (Lyall et al., 2009; Organization for Economic Cooperation and Development-OECD, 2015; Bermúdez et al., 2016). Similarly, there is a lack of studies that incorporate the omnipresent characteristics of the group's human relationships and common history and its influence in the research enterprise (Tannenbaum et al., 2012). This is not an easy task: to combine a useful model for predictions in the research fields with the challenges of the context.

In this chapter, we propose the use of a heuristic methodology, together with the methodology for the investigation of case studies based on (Aamodt & Plaza, 1994), for the identification of interdisciplinary research subjects. Using a historical-critical technique, a data set of research projects of the group in question (in a given period of time) is used to extract the research topics that have been treated recurrently. The most relevant factors that allow characterizing the identified research topics, with a reductionist method, are related as follows: (1) level of learning acquired by the group (mastering of the research field), (2) academic quality achieved in the related scientific production, and (3) its possible contribution to the generation of new strategic knowledge (in the mark of development objectives).

13.3 Methods

Case study. Context of application and description of the research group: The "Research Group in Technological and Environmental Developments" (here RG-TED, GIDTA for its Spanish acronym) is an interdisciplinary research team in the field of engineering, particularly environmental, telecommunications, and industrial engineering. The group is part of the Faculty of Engineering and Architecture of the Catholic University of Manizales, a 64-year-old, small (approx. 4000 students), private and non-

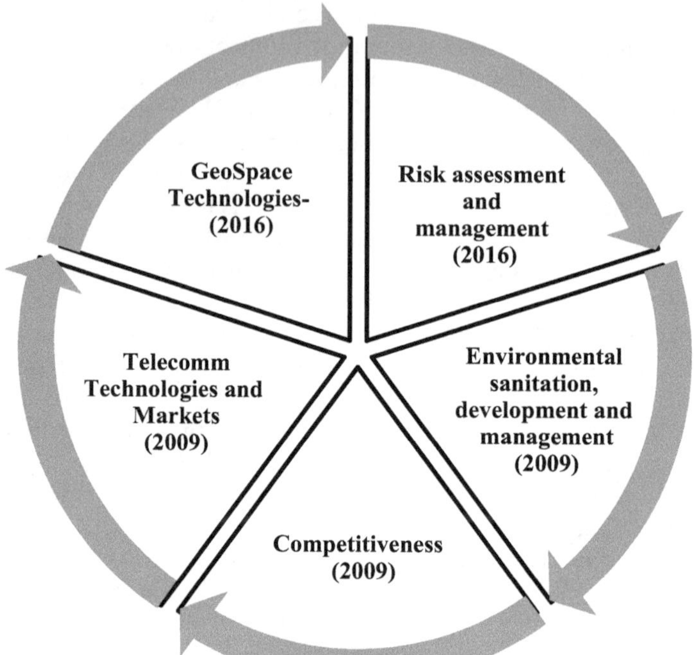

Fig. 13.2 RG-TED research lines and its year of creation

profit university in Caldas, Colombia. Five research lines are declared by RG-TED (Fig. 13.2). Within these lines of research, particular research topics have emerged throughout the years in a request not only to local context needs and opportunities but also as particular interests or circumstances of the research staff. Some of these topics are shown in Fig. 13.3.

The Administrative Department of Science, Technology, and Innovation of Colombia (Colciencias) is a governmental entity responsible for the application of the public policies of science, technology, and innovation, and to define and measure the quality standards of the researchers and research groups in the country. Colciencias has ranked RG-TED as type C group, in a four-scale metrics (from top to bottom A1, A, B, and C).

Record of projects and indicators: a list of institutionalized projects in the period 2001–2013 was taken from the public Colciencias "GrupLac" platform (online). For the projects carried out in the years 2014 to 2016,

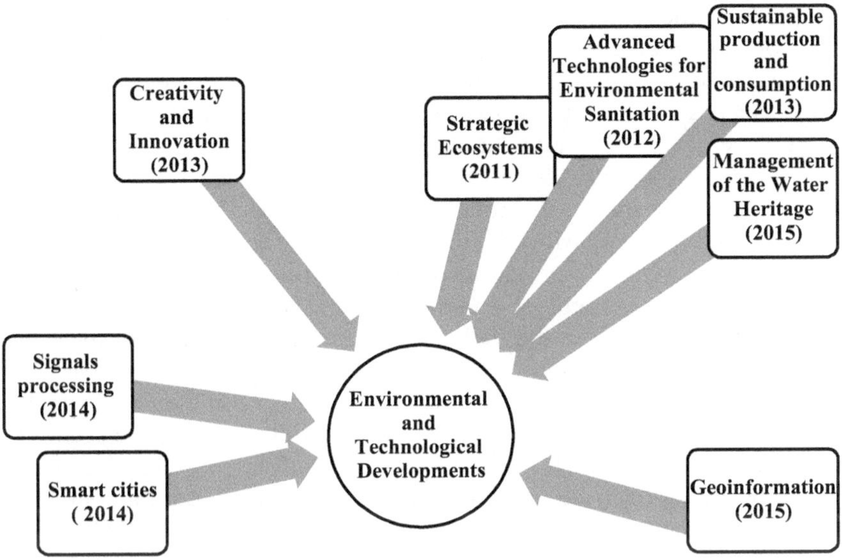

Fig. 13.3 Particular research topics emerged from the RG-TED research lines

the lists of approved projects were published by the research office of the university (the complete list of identified projects can be delivered on request). A total of 59 research projects were accounted in the period 2001–2016, of which 21 of them were approved in 2001–2013, 20 in 2014, 9 in 2015, and 9 in 2016. The main research topic in each project was identified based on its title and, in most cases, with the help of prior knowledge of the project at hand. It is clear that a project is developed around a topic and that different projects may be related to the same topic. Considering the list of projects, six indicators were proposed as follows:

- Continuity Indicator: It measures the continuity in the time a research topic has been worked on. The total period (2001–2016) was subdivided into four sub-periods, as follows: first period 2001–2013, second period 2014, third period 2015, and fourth period 2016. This indicator measures in a normalized scale between 0 and 1, the fraction of total time that the topic has been worked on, taking values of at least 0.25 for a topic that has only been worked in one of the four periods and a maximum of 1.0 for a subject that has been worked continuously in the four periods.

- Incidence Indicator: It measures the incidence (frequency) with which a research topic has been worked on. The total number of projects carried out on the subject was counted in the total observation period 2001–2016. To normalize the value of the indicator between 0 and 1, the total number of projects in the theme was divided by the maximum value found for a topic.
- Interdisciplinarity Indicator: It measures the level of interdisciplinarity in the treatment of a research topic. A scale of seven grades is proposed based on the basic definitions of multidisciplinarity, interdisciplinarity, and transdisciplinarity. The scale is shown in Table 13.2.
- Complexity Indicator: It measures the level of complexity in which a research topic has been worked on. For this purpose, the classification of the research products for publication, proposed inside the group, has been adopted. The complexity is measured indirectly in a scale of contribution to the area of knowledge as follows: the lower levels of the scale are smaller contributions to the field and imply lower degrees of intellectual elaboration than those in higher levels of the scale. The proposed scale and the value of the related indicator are presented in Table 13.3. The scale is intended to capture the grade of complexity not only of the published papers and books (research products which are measured with classical indicators such as h-index, the impact factor of the scientific journal, among others) but also in the achieved goals and results of every analyzed project.

Table 13.2 Level of interdisciplinary treatment of a research topic and interdisciplinarity indicator

Level of interdisciplinary	*Scale*	*Interdisciplinarity indicator*
Multidisciplinary (without complex interplay)	Null	0
Multidisciplinary with very low interplay	Very low	0.1
Multidisciplinary with low interplay	Low	0.25
Multidisciplinary with high and complex interplay	High	0.5
Multidisciplinary with highly complex interplay	Very high	0.75
Full spectrum of related disciplines involved and interconnected	Complete	1
Transdisciplinary (beyond interdisciplinary boundaries)	NA	NA

Table 13.3 Contribution scale of a research topic to the area of knowledge and complexity indicator

Contribution to the area of knowledge of a research topic	*Scale*	*Complexity indicator*
Literature review	1	0.13
Emulation of methodology with new results	2	0.25
New experimental/empirical evidence	3	0.38
Application of known methodologies to new case studies	4	0.5
Development and evaluation of new methodologies	5	0.63
Critical treatment of a field of knowledge with interpretative contributions	6	0.75
New technological applications with proof of concept or field evaluation	7	0.88
Presentation of discoveries with experimental evidence	8	1

Table 13.4 Prioritized sustainable development goals for RG-TED/context and congruence indicator

Sustainable development goals		*Congruence indicator*
Clean water and sanitation	A	1
Affordable and clean energy	B	0.8
Sustainable cities and communities	C	0.7
Responsible production and consumption	D	0.5
Climate action	E	0.3
Life on land	F	0.3
Industry, innovation, and infrastructure	G	0.3

- Congruence Indicator: It measures the level of congruence (to an external reference) that a research topic has been worked on. Seven of the 17 Sustainable Development Goals (SDGs) were adopted (which had been previously suggested inside the group's management) as the most related to the research lines declared. These SDGs were associated with the identified themes and ranked in order of the perceived priority for the local region. A standardized indicator is shown in Table 13.4. In general, all the SDGs are equally important and their prioritization is circumstantially assigned in this work. That is why none SDGs received a null value indicator, and for the three SDGs with less priority, the same score was assigned: 0.3 points.

- Global Projection Indicator for the Research Topic: It measures the level of total projection (strategic value for present and future actions) of a research topic that has been worked on. The values of continuity, incidence, interdisciplinarity, complexity, and congruence indicators were added. The value of the indicator was normalized dividing the result by the highest value obtained. As a result, the comparison was simpler and referred to the project with the highest value of the global indicator.

Identification of a research framework with external partners and stakeholders: In order to look for strategies to improve the performance of the STI Colombian model, in this work we look at two main actors and their structural coupling in the STI system: academic research groups (RG-TED in this case) and productive sector (firms and industries within the local context). Using case-based reasoning, the team members analyzed the projects of the research group, looking for patterns that have appeared in the interactions with external partners and stakeholders. After several meetings, four key factors of the interaction were identified (focus, time, money, and trust) and the main characteristics of each one were established according to the peculiarities of the social and cultural context of the research actions. Based on this approach, a set of intervention strategies is proposed to directly attack the reasons that make the structural couplings so stiff.

13.4 Results

Identification of interdisciplinary research subjects: Among the 59 projects worked on by RG-TED in the period 2001–2016, it was possible to extract a total of 22 differentiable research topics. The results obtained for the different topics and related indicators are presented in Table 13.5. The topics ranked by their global projection indicator are presented in Fig. 13.4. For the 22 research topics that were extracted from the list of 59 institutionalized projects, topics of the environmental research line appear more frequently (with a greater number of projects), and the industrial and telecommunications topics less frequently. The topics in risk management and geospatial technologies research lines are minor due to their recent incorporation to the group interests in 2016.

Regarding the continuity indicator, the topics that have had the greatest continuity are conservation of high Andean wetlands, environmental

Table 13.5 RG-TED historic research subjects and indicators for knowledge management

Research subject	*Number of projects developed*				*Continuity and incidence*			*Interdisciplinarity*		*Complexity*		*Congruence*		*Global projection*	
	2001–2013	*2014*	*2015*	*2016*	*COI*	*Total*	*INI*	*Scale*	*IDI*	*Scale*	*CXI*	*Type*	*CGI*	*Total*	*GPI*
Biotechnology	1	0	0	0	0.25	1	0.2	Null	0	2	0.25	G	0.3	0.97	0.25
Ecology/sustainability of bamboo forests	3	0	0	0	0.25	3	0.5	Low	0.25	5	0.63	F	0.3	1.93	0.49
Wastewater treatment	3	0	0	0	0.25	3	0.5	Null	0	5	0.63	A	1	2.38	0.61
Remote laboratories	2	0	0	0	0.25	2	0.3	Null	0	7	0.88	G	0.3	1.76	0.45
Artificial wetlands	0	2	1	1	0.75	4	0.7	Very low	0.1	2	0.25	A	1	2.77	0.7
Waste valorization	0	3	1	1	0.75	5	0.8	Null	0	2	0.25	D	0.5	2.33	0.59
Cleaner production in agriculture	1	1	0	0	0.5	2	0.3	Low	0.25	2	0.25	D	0.5	1.83	0.47
Biodiversity/social appropriation of knowledge	0	1	0	0	0.25	1	0.2	Very high	0.75	5	0.63	F	0.3	2.09	0.53
Bioprospection/phytoremediation	0	2	0	0	0.25	2	0.3	Low	0.25	2	0.25	A	1	2.08	0.53
Treatment of landfill leachate/waste valorization	1	1	0	1	0.75	3	0.5	High	0.5	2	0.25	A	1	3	0.76
Conservation of high Andean wetlands	3	1	1	1	1	6	1	High	0.5	5	0.63	F	0.3	3.43	0.87
Environmental management systems	2	1	1	1	1	5	0.8	Null	0	5	0.63	D	0.5	2.96	0.75
Education for sustainable consumption	0	2	0	0	0.25	2	0.3	Low	0.25	5	0.63	D	0.5	1.96	0.5
Optimization of small and medium enterprises operation	2	1	2	1	1	6	1	Null	0	5	0.63	G	0.3	2.93	0.75
Business innovation in small and medium enterprises	0	0	1	0	0.25	1	0.2	High	0.5	5	0.63	G	0.3	1.84	0.47
Process simulation	0	1	0	0	0.25	1	0.2	Null	0	5	0.63	G	0.3	1.34	0.34
Software development	0	1	0	0	0.25	1	0.2	Null	0	7	0.88	G	0.3	1.59	0.41
Communication system/Digital signal processing	1	1	1	0	0.75	3	0.5	Low	0.25	7	0.88	G	0.3	2.68	0.68
Environmental telemetry	2	1	1	1	1	6	1	Very high	0.75	7	0.88	F	0.3	3.93	1
Remote sensing/telemetry/wetlands	0	1	1	0	0.5	2	0.3	Very high	0.75	8	1	F	0.3	2.88	0.73
Remote sensing/risk management	0	0	0	1	0.25	1	0.2	Very high	0.75	4	0.5	C	0.7	2.37	0.6
Remote sensing/image processing	0	0	0	1	0.25	1	0.2	Very low	0.1	5	0.63	G	0.3	1.44	0.37

COI Continuity indicator, *INI* incidence indicator, *IDI* interdisciplinarity indicator, *CXI* complexity indicator, *CGI* congruence indicator, *GPI* global projection indicator

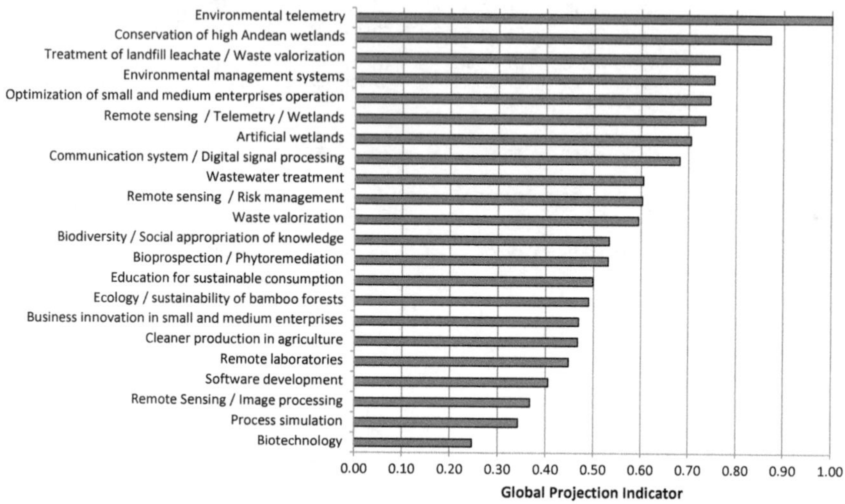

Fig. 13.4 RG-TED historic research subjects ordered by their global projection indicator

management systems, environmental telemetry, and process optimization in small- and medium-sized enterprises (SMEs) with up to 100% continuity in the period. It was followed by the topics of artificial wetlands, waste valorization, treatment of landfill leachate/waste valorization, and communication systems with digital signal processing (75% continuity).

According to the incidence indicator, a greater number of projects have been presented within the topics of high Andean wetlands, optimization of processes in SMEs, and environmental telemetry, followed to a lesser extent by the topics of artificial wetlands, waste valorization, and environmental management systems. It is interesting to note that the topic of environmental management systems does not seem to be more frequently related to the topics of higher incidence in research, even its enormous importance for sustainability in the territory.

According to the interdisciplinarity indicator (an index of cohesion that reflects some type of interdisciplinary work between the lines), the results were 36.4% of the topics correspond to the null scale, 9.1% to very low scale, 22.7% to low scale, 13.6% to high scale, and 18.2% to very high scale. Therefore, the main of the research work of RG-TED is up-to-now almost of a single disciplinary nature, which must be improved. Among the most

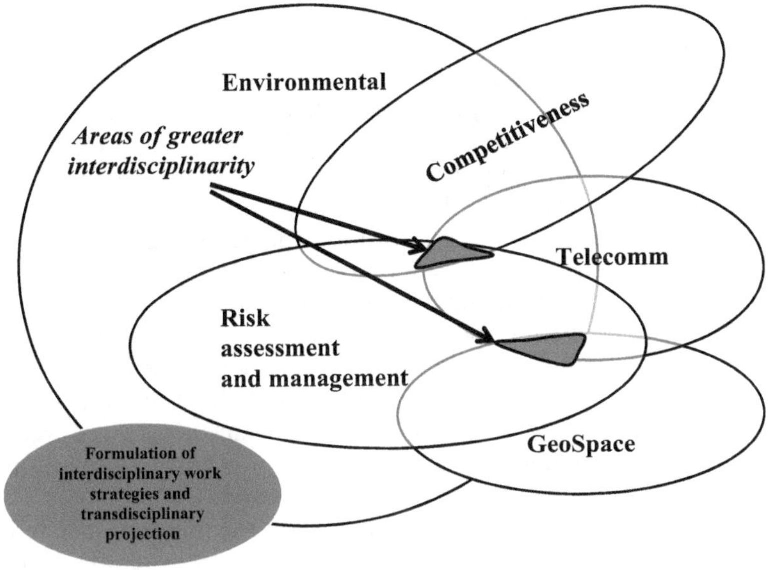

Fig. 13.5 Conceptual map for the emergence of interdisciplinarity in RG-TED

interdisciplinary projects are environmental telemetry, remote sensing/telemetry/wetlands, remote sensing/risk management, and biodiversity/social appropriation of knowledge. Those projects agree with the context of a big agricultural territory: full of great biodiversity, prone to natural disasters, able to adopt and adapt advanced technologies to enhance conservation, production, and to prevent disasters. Fig. 13.5 illustrates a conceptual map for the emergence of interdisciplinarity in RG-TED.

It must be taken into account that, although they were analyzed as different subjects, the topics 5 (artificial wetlands) and 9 (bioprospecting for phytoremediation) (Table 13.5) are closely related, since the microbiota of plants has been studied for phytoremediation processes. However, in RG-TED, both topics have been oriented by different researchers and without cohesion. Considering both issues together, their interdisciplinarity indicator (and overall projection indicator) could be among the highest.

In relation to the complexity indicator, the subjects that contribute the most to the area of knowledge are remote sensing and telemetry for high Andean wetlands (research approach with maximum level of contribution to the field), followed to a lesser extent by the following topics: remote laboratories, software development, communication systems with digital signal processing and environmental telemetry. With a smaller contribution to the areas of knowledge, different projects have been developed in the following topics: ecology and sustainability of bamboo forests, wastewater treatment, high Andean wetlands, environmental management systems, sustainable consumption education, operation optimization in SMEs, process simulation and SMEs' business innovation. It must be pointed out that the classification obeys to the perceived contribution to the area of knowledge of the related projects.

The interpretation of the value of the indicator of congruence with the sustainable development goals is more difficult. For example, if the goal of clean water and sanitation is taken as a higher priority, the directly related subject of wastewater treatment turns out to be one of the ones with low continuity, incidence, and complexity. The themes of artificial wetlands and treatment of landfill leachates with natural coagulants stand out in relation to the priority of the clean water objective.

Finally, Fig. 13.4 shows that the topics with the greatest value in their global projection indicator are environmental telemetry, conservation of high Andean wetlands, treatment of landfill leachate/waste valorization, environmental management systems, optimization of operations in SMEs, and remote sensing/telemetry/wetlands. These results guide the group to establish an interdisciplinary macro project considering a research program in competitive sustainability of the Andean basin territory. Note that this macro project could point out to the use of remote sensing and environmental telemetry as tools to create specially designed processes of environmental management in the local territory, including the optimization of operations in agro-industrial SMEs and the conservation and responsible use of strategic ecosystems of the territory such as the high Andean wetlands. Some contributions of RG-TED to this goal are presented in Holguin et al. (2016) and Flórez, Rincón, Cardona, and Alzate (2017).

Identification of a research framework with external partners and stakeholders: Although the proposed methodology is an interesting tool for the selection of strategic themes for research considering different indicators such as continuity, incidence, complexity, and congruence, these are not the only aspects to consider. The external partners and stakeholders are very important because they are part of the context and most of the time decide and/or affect the course of research actions. Figure 13.6 presents an onion-type conceptual frame to approach the relation between the research activity performed by the researcher, the general academic factors

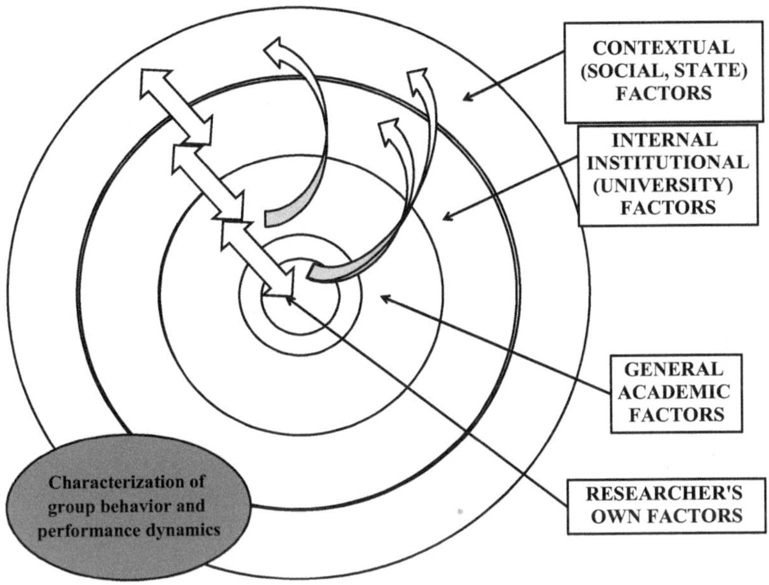

Fig. 13.6 Conceptual frame in order to understand the group behavior and performance dynamics

(such as the advance and frontier of knowledge), the institutional factors (such as research policy and budget), and the contextual factor (such as governmental research policies and metrics of research performance). There are many interactions among the shields and everyone must be analyzed in order to understand the group behavior and performance dynamics.

Four key factors of the interaction were identified (focus, time, money, and trust) and the main characteristics of each one were established, as shown in Table 13.6. The analysis of the peculiarities of the social and cultural context of the research actions showed that soft competencies, rather than hard (technical-scientific) competencies, were key to successful or unsuccessful interactions with partners and stakeholders. Based on this fact, a set of intervention strategies is proposed as shown in Table 13.6.

As a synthesis of the proposed methodology and obtained results in the case of RG-TED, a general strategy for the knowledge management for research groups in LA&C countries can be formulated, as shown schematically in Fig. 13.7. Research is underway to generate more insights and understandings about the most effective and appropriate models to be applied in these territories.

Table 13.6 Key factors of the interaction in external couplings for RG-TED

Issue	*Higher educational institutions (research groups)*	*Productive sector*	*Intervention strategy*
Focus	Basic research Contribution to science and knowledge	Applied research The need to solve their problems. They look for profit margin	Both actors must know explicitly and consider the expectations of the other and visualize a common goal beyond their particular interest (like community development)
Time	Slow Projects to long term	Fast Projects to short term	Execution time for a research project must be considered as a resource and must be optimized using other resources like more research staff.
Money	Low Budget, need to find co-financers	They have money, but R&D&I are not their priority	Apply to national and international governmental and NGOs grants
Trust	Both think that the other counterpart will take advantage of the other		The most important factor. Always approach the interaction with human values such as honesty, humility, and respect

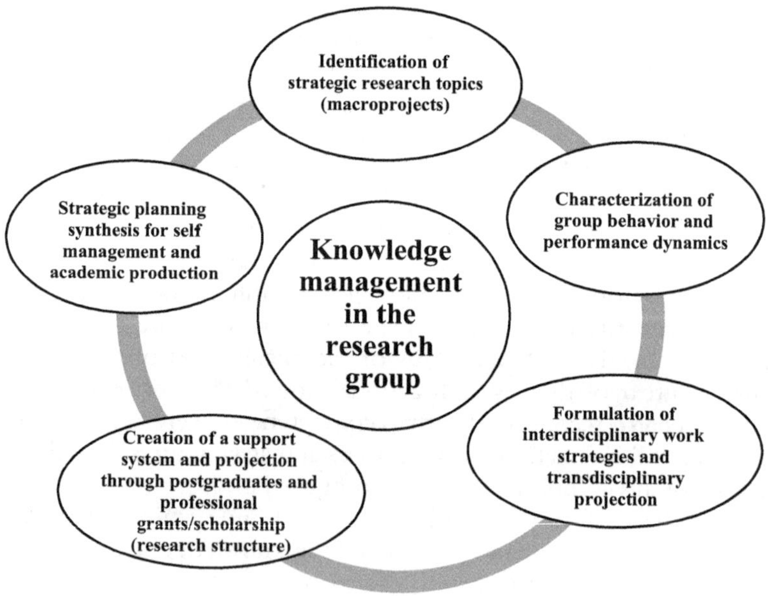

Fig. 13.7 A general strategy for the knowledge management for research groups in LA&C countries

13.5 Conclusions and Significance

The proposed way of thinking considers general ideas about the realities of development in emerging economies (some of them proven as true or useful), and conceives the scientific research as a human endeavor in order to urgently increase human well-being in developing countries. The use of the historical particularities of the research group, as a valuable source of new interpretations of reality, is presented for a "change of mentality", considering action at the individual and community levels.

The scope of this chapter was to take a closer look at the scientific research activities that a small research group from Colombia has done in the field of engineering since 2010, in order to improve our knowledge management and to get a positive impact in our society. We cannot provide an authoritative view of these matters, because our low score in the national science rankings (as many of the 700 research groups in Colombian), and because we do not have international recognition yet. It is our desire that the perspectives grounded in our own experience and presented in this chapter improve over that situation and also can be useful for small, young, and enthusiastic research groups in the developing countries in LA&C and in other parts of the world. Similarly, it can provide them with some tools that can be adapted to its own social context, to accomplish its goals of general and global contributions of science and technology to human knowledge, well-being, and development for all.

The application of the proposed methodology to this case study allowed to identify a group of strategic macro projects of an interdisciplinary nature from the group's work history, the knowledge acquired and developed, and the needs of the regional context. The general aspects of the strategy were discussed, and some recommendations for its adaptation to other case studies and other areas of research are carry out a debate about the results of this study and its possible interpretations within your own research group in a debate format. Moreover, this study could be improved by more and better indicators designed to capture the dynamic landscape of research and human development.

References

Aamodt, A., & Plaza, E. (1994). Case-Based Reasoning: Foundational Issues, Methodological Variations, and System Approaches. *Artificial Intelligence Communications, 7*(1), 39–59. IOS Press. Retrieved from https://www.idi.ntnu.no/~agnar/publications/aicom-94.pdf

Ahmadpoor, M., & Jones, B. F. (2017). The Dual Frontier: Patented Inventions and Prior Scientific Advance. *Science, 357*(6351), 583–587. https://doi.org/10.1126/science.aam9527

Bergek, A., Jacobsson, S., Carlsson, B., Lindmark, S., & Rickne, A. (2008). Analyzing the Functional Dynamics of Technological Innovation Systems: A Scheme of Analysis. *Research Policy, 37*(3), 407–429. https://doi.org/10.1016/j.respol.2007.12.003

Bergek, A., Hekkert, M., Jacobssonc, S., Markardd, J., Sandénc, B., & Truffer, B. (2015). Technological Innovation Systems in Contexts: Conceptualizing Contextual Structures and Interaction Dynamics. *Environmental Innovation and Societal Transitions, 16*, 51–64. https://doi.org/10.1016/j.eist.2015.07.003

Bermúdez, L., Moreno, E., & Rios, M. (2016). Innovation Strategies Developed by Research Centers in Public Universities of the Department of La Guajira, Colombia. *INGE CUC, 12*(1), 32–41.

Consejo Nacional de Política Económica y Social-CONPES, Dirección Nacional de Planeación de Colombia-DNP. (2015). *Política Nacional de Ciencia, Tecnología e Innovación 2015–2025.* Borrador. Retrieved from http://www.colciencias.gov.co/sites/default/files/upload/noticias/conpes-borrador-cti.pdf

Cornell University, INSEAD, and WIPO. (2017). *The Global Innovation Index 2017: Innovation Feeding the World.* Retrieved from http://www.wipo.int/edocs/pubdocs/en/wipo_pub_gii_2017.pdf

Fairbanks, M., & Lindsay, S. (1997). *Plowing the Sea: Nurturing the Hidden Sources of Growth in the Developing World.* USA: Harvard Business Review Press.

Flórez, G., Rincón, A., Cardona, P., & Alzate, A. (2017). Análisis multitemporal de las coberturas vegetales en el área de influencia de las minas de oro ubicadas en la parte alta del sector de Maltería en Manizales, Colombia. *Dyna, 84*(201), 95–101. https://doi.org/10.15446/dyna.v84n201.55759

Frodeman, R. (Ed.). (2017). *The Oxford Handbook of Interdisciplinarity* (2nd ed.). UK: Oxford University Press.

Fortunato, S., Bergstrom, C., Börner, K., Evans, J., Helbing, D., Milojević, S., … Barabási, A. (2018). Science of science. *Science, 359*(6379). https://doi.org/10.1126/science.aao0185

Gurevitch, J., Koricheva, J., Nakagawa, S., & Stewart, G. (2018). Meta-analysis and the Science of Research Synthesis. *Nature, 555*, 175–182. Retrieved from https://www.nature.com/articles/nature25753

Gosens, J., Lu, Y., & Coenen, L. (2015). The Role of Transnational Dimensions in Emerging Economy 'Technological Innovation Systems' for Clean-tech. *Journal of Cleaner Production, 86*(1), 378–388. https://doi.org/10.1016/j.jclepro.2014.08.029

Gray, S., Zanre, E., & Gray, S. (2013). Fuzzy Cognitive Maps as Representations of Mental Models and Group Beliefs: Theoretical and Technical Issues. In E. Papageorgiou (Ed.), *Fuzzy Cognitive Maps for Applied Sciences and Engineering – From Fundamentals to Extensions and Learning Algorithms* (pp. 29–48). Berlin: Springer Publishing.

Holguín, M., Cardona, O., Sierra, E., Mejía, J., Orozco, A., & Castellanos, G. (2016). Machine Fault Detection Based on Filter Bank Similarity Features Using Acoustic and Vibration Analysis. *Mathematical Problems in Engineering., 2016*, 1–14. https://doi.org/10.1155/2016/7906834

Henly-Shepard, S., Gray, S., & Cox, L. (2015). The Use of Participatory Modeling to Promote Social Learning and Facilitate Community Disaster Planning. *Environmental Science & Policy, 45*, 109–122. https://doi.org/10.1016/j.envsci.2014.10.004

Lyall, C., Williams, R., & Meagher, L. (2009). *A Short Guide to Developing Interdisciplinary Strategies for Research Groups*. Number 7: Developing Interdisciplinary Strategies for Research Groups. ISSTI The University of Edinburgh. Retrieved from https://www.researchgate.net/publication/265425942_A_Short_Guide_to_Developing_Interdisciplinary_Strategies_for_Research_Groups

Organización para la Cooperación y el Desarrollo Económico. (2015). *Frascati Manual 2015: Guidelines for Collecting and Reporting Data on Research and Experimental Development, The Measurement of Scientific, Technological and Innovation Activities*. Paris, Francia: OECD Publishing.

Pentland, A. (2012). *The New Science of Building Great Teams*. USA: Harvard Business Review.

Tannenbaum, S., Mathieu, J., Salas, E., & Cohen, D. (2012). Teams Are Changing: Are Research and Practice Evolving Fast Enough? *Industrial and Organizational Psychology, 5*(1), 2–24. https://doi.org/10.1111/j.1754-9434.2011.01396.x

United Nations Development Programme. (2016). *Human Development Report 2016: Human Development for Everyone*. Retrieved from http://hdr.undp.org/sites/default/files/2016_human_development_report.pdf

Zeng, A., Shen, Z., Zhou, J., Wu, J., Fan, Y., Wang, Y., & Stanley, H. (2017). The Science of Science: From the Perspective of Complex Systems. *Physics Reports, 714–715*, 1–73. https://doi.org/10.1016/j.physrep.2017.10.001

CHAPTER 14

An Analysis of Urban Public Policy Regarding Science, Technology and Innovation from the Perceptions of Stakeholders: A Case Study for a City of an Emerging Country

Alexander Cotte Poveda, Clara Carolina Jimenez, and Clara Inés Pardo Martínez

14.1 Introduction

In the past decade, science, technology and innovation (STI) have become more relevant, starting as a marginal term inside States and becoming important as a transversal policy in some countries as a way to integrate different sectors and areas. The Organisation for Economic Co-operation and Development (OECD) says that the science, technology and innovation

A. Cotte Poveda
Universidad Santo Tomás, Bogotá, Colombia

Colombian Observatory of Science and Technology (OCyT), Bogotá, Colombia

C. C. Jimenez
Colombian Observatory of Science and Technology (OCyT), Bogotá, Colombia

C. I. Pardo Martínez et al. (eds.), *Analysis of Science, Technology, and Innovation in Emerging Economies*,
https://doi.org/10.1007/978-3-030-13578-2_14

strategy in different countries is focused on using these tools to find new resources for growth and competition, looking for any way to restore competitive industries by investments in innovation and industrial projects. An example of that type of country is France with its programme "Investissements d'avenir", PIA (OECD, 2012).

On the other hand, some States have tried to develop a new industrial policy by focusing on a technology strategy, applying nanotechnologies, biotechnologies and communication technologies to traditional sectors, such as agriculture. Brazil and China are proponents of this type of policy (OECD, 2012). Focusing on different perspectives, a new industry policy related to climate change, energy safety and food safety, among others, is prevalent in Italy, Japan and Germany. Another strategy, focusing on stability in development and investment, is a controlled policy that emphasizes fiscal policy and control of expenditures (OECD, 2012).

The need for innovation policies has always existed in the world, and it is necessary for a country to maintain the capacity to innovate because tools of innovation are necessary for a country to be competitive. We cannot ignore the importance of the participation of the population. A policy without approval of the population cannot be successful, as approval is related to preservation, retention and actualization of human talent, and the OECD says that this is as essential as State support is to business.

In Latin America, different countries have developed a stable STI strategy and a durable public policy. An example of this is Costa Rica, which has developed a national strategy to involve society in science, technology and innovation but has focused on enterprises and productivity. However, the strategy also includes education, environment, the energy matrix, health, agriculture and convergent technologies (Science, Technology and Innovation Ministry, 2015).

Ecuador has implemented an information-communication technology (ICT) policy that is transversal, and it focuses on new knowledge efficiency. The effectiveness of the policy has been shown to depend on the State's solid scientific infrastructure and economic support, in which education and knowledge transference are central to the search for better human development (Sela, 2016a, b).

C. I. Pardo Martínez (✉)
School of Management, Universidad del Rosario, Bogotá, Colombia
e-mail: clara.pardo@urosario.edu.co

Innovation is the policy focus in Chile, which declared that it is possible that a country's increased growth, propelled by innovation, requires a shared vision among human capital, innovation and science. To do this requires coherence, definition of strategy objectives, inclusion of stakeholders, university education for research on the subjects of industry and enterprise, and transfers to the economic world (National Council of innovation for competitiveness, 2012).

The policies mentioned before are new, as none have existed for more than three years, but the Colombian policy was started in 2009, with the Conpes 3582 policy. This policy attempted to focus entirely on STI; however, by formally committing to many entities for the implementation of the policy, disparities and ruptures in development were generated because the role of each entity within the agreed instruments and strategies was not clear. This is a common element in District Policy: Science, Innovation and Technology Bogota 2007–2019.

Cities of the world are striving to become smart and sustainable by creating a relationship between a science, technology and innovation policy and the city. More than half of the world's population lives in urban and peri-urban zones, creating a challenge for cities in developmental and economic terms, as population growth is increasing more every day. Because of this, it is necessary to have cities that are planned, constructed and administered to provide sustainable benefits. Technology must be a facilitating mechanism to monitor and provide solutions for the needs of citizens. Science can provide input to solve cohesion and social problems, spatial development, urban frameworks, urban economics, urban ecology and the environment, among others. Finally, innovation provides a fundamental tool for application to all the above-mentioned issues (Guttieres, Fransiska, & Kusuma Dewi, 2016).

Population growth is not the problem; it is the issues associated with population, such as infrastructure deficit, resource depletion, environmental deterioration, risks and natural disasters. Public problems exist that government must solve, beginning with public policies that generate more welfare for citizens. STI has helped different cities to overcome public problems by generating geospatial tools for spatial planning, helping to solve transportation difficulties and promoting economic and environmental tools such as the use of bicycles, walking or promoting high-speed public transport as the main means of individual transportation.

Another important STI tool is the mobile app that has the capability to report a natural disaster in advance or a traffic situation. The monitoring of resource consumption has begun, generating economic incentives, for example, for promoting responsible and measured consumption (Commission on Science and Technology for Development, 2013).

Urban planning implies that conditions across cities vary broadly, and the priorities and requirements for both analysis and interventions demand to be supported in the specificity of spaces, where the application of urban policies could combine distributed sources of knowledge and viewpoints on urban living or interest groups, where STI should put in the service of social improvement (McFarlane & Söderström, 2017). In this context, this study seeks to analyse the main elements of STI policy for Bogotá from the viewpoints of stakeholders as a contribution and input to build STI policies in urban contexts.

From this background, the main goals of this study are twofold. First, the study seeks to analyse the main perceptions of stakeholders with respect to the formulation of STI policy and the main STI trends in Bogotá. Second, using different techniques to establish the main elements for consideration in the policy, the study also seeks to determine the main factors that could influence an adequate STI policy for Bogotá, from the perspective of interest groups. The research question that guides this study is as follows: What are the main factors that STI policy should include, from the perspective and opinion of the main interest groups in Bogotá? Studies in this area are limited, and research in Latin America is beginning in order to find a strategy for urban sustainable development; this study is key for policymakers to develop and design policies and instruments that enhance the importance of STI in developing countries.

Thus, this study provides a more accurate description of the situation and perceptions of STI in urban areas such as Bogotá, which represents cities of developing countries, contributes to the literature on STI policy and provides novel possibilities for the integration of STI into adequate policies in an urban context. The study also contributes to the limited analysis of the importance of STI to promote sustainable cities and the well-being of the urban population.

From those elements, this chapter is structured as follows. First, this introduction is presented. Second, the science, technology and innovation policy trends for Bogotá in recent years are presented. Third, the methods used in this study are described. The results are presented in Sect. 14.4. Section 14.5 shows a discussion of the main results. Finally, we present a series of conclusions and recommendations resulting from this study.

14.2 Trends in Science, Technology and Innovation (STI) Policy in Bogotá

In 2007, Bogotá defined its public policy on STI planned through 2019. In the policy, it is possible to identify a restatement of the national STI policy, that is, to increase the capacity of the country to identify, produce, disseminate, use and integrate scientific and technological knowledge, with the aim of improving competitiveness and contributing to the country's productive transformation (Conpes, 2009), as a general objective. Bogotá's policy is stated as follows: Define strategic programmes and projects to consolidate Bogotá as a knowledge society in which scientific capacities, technological development and innovation contribute to the social and economic development of the city, with criteria of inclusion and equity (District Commission for science, technology and innovation, 2007). We can see many differences:

- First, in the country situation, the aim is to generate and produce more knowledge; Bogotá seeks to define strategic programmes and projects that do not necessarily include more knowledge.
- Second, it is possible to see a conflict of interests in the scientific committee of the District, which is made up of a large number of institutions, such as the different District Offices, Bogota Telecommunications Company, Bogota Energy Company, José Celestino Mutis Botanical Garden, Colombian Institute for Development Science and Technology—COLCIENCIAS, SENA, Live Systems Technology SA, Bogota Chamber of Commerce, IT Group Deusto Colombia SA and Maloka Corporation. Conflict is possible since for many of the institutions, their interests and their contributions are already included within the different District Secretariats. In addition, it is important to consider that these are institutions that can be allied with politics and not necessarily in conformance with the scientific committee. The need to make this policy a participatory one is understood, but this does not mean that there should be participation by each entity within the scientific committee, especially due to the diverse interests of these different institutions.
- Objectives are not clear, and they are difficult to size. In addition to the existence of a disconnect between the conceptual framework, the principles and the plans, the most obvious example is the approach within the framework on equality and equity, an element reflected

neither in the plans nor in the programmes, reflecting a disconnect in the policy. Similar disconnects between the plans and the programmes are evident, because in attempting to cover all the STI terms, there are projects that have not been started, such as the creation of a science and technology programme for physical activity and sport, among others.

Table 14.1 shows the current composition of the district STI policy as planned for the period 2007–2019, describing mission and vision, objectives and scope, and four core ideas with respective programmes. However, this policy is limited by current requirements to strengthen STI in urban areas, in which this study aims to provide analytical tools and recommendations for the formulation of the 2017 District Science, Technology and Innovation Policy. However, this plan should also include the following elements that are key to the STI policy for Bogotá:

Definition of specific objectives: Quantify objectives to promote an STI culture in the region; encourage innovation programmes inside the region open to all citizens; share and generate information, convocations, programmes and accountability open to all stakeholders; and identify emerging sectors in STI, among other related aspects.

Do not attempt to cover all areas: It can be promoted more as a transversal policy. STI has to be a geographically focused term that promotes the use of technology, knowledge transfer and innovation in all sectors. Do not implement a programme for a specific sector, such as an "STI programme for sports". It would be better to have active participation of these sectors in STI formulation, such as a discussion about the need for innovation for the promotion of physical activity in sports in the District, for example, on the transfer of knowledge and use of technology in the health sector.

If some programmes and observatories are not working, it is better to limit programmes to ones that cover the most important points and that can generate the greatest impact: The development of programmes is essential for public policy, but they cannot be designed for only one area. To be relevant and impactful, programmes promoted and developed must fuel policy; organize and promote the participation of knowledge networks; identify, highlight and replicate successful innovation and technology in the region; and generate development and well-being, for example, the Innovation Consultancy in Universities programme and STI programmes that contribute to the peace process.

Table 14.1 The current scheme of the STI policy in Bogotá

Mission: Promotion and institutionalization of knowledge in the region, through the strengthening of scientific, technological and innovation activities, based on the consolidation of the scientific community, entrepreneurial and innovative citizens, and the community in general.
Vision: In 2019, Bogotá DC is recognized in Colombia and in the world as a knowledge society and innovation economy, based on education, scientific and technological culture, social appropriation of knowledge, transfer of advanced technologies and innovation, with an elevated level of population's life quality.

Scope	Objectives
• Synergy between public and private actors for the development of innovative productive activities. Making up the Science, Technology and Innovation District System and incorporating innovation and technological development.	• Define strategic programmes and projects to consolidate Bogotá as a knowledge society in which scientific capabilities, technological development and innovation contribute to the city's social and economic development, with inclusion and equity as rules.
• This plan is articulated with a set of public policies and private initiatives that are advanced in the Capital District, and at the regional and national levels.	• Boost institutionalism of the work, technology, and innovation in science in the city through coordination, articulation, dissemination and financing of the plan.
• Replying to social, productive and environmental needs of the Capital District, linking and fostering scientific and technological supply of different public and private institutions.	• Promote research, innovation and technological development, which reflect social and environmental requests for increasing quality of life for women and men of the city.
• Covering various fields of scientific and technological development, such as basic research, applied research, experimental development, innovation, design, technological services, technical assistance and training. Support political and technical decisions that promote social and economic development of the city.	• Encourage formation and promotion of the scientific spirit, technological and innovative thought and capacities of the population for the growth of the scientific and technological culture.
	• Promote sustainable technological development, innovation and modernization of productive activities, in pursuit of economic growth related to the welfare of the society who lives in the city.

(*continued*)

Table 14.1 (continued)

Core ideas	Programmes
Research, innovation and technological development	• Research programme on the social, economic, cultural, political and environmental determinants of the living conditions of the population. • Science and Technology Program for Physical Activity and Sport.
Training and promotion of the scientific, innovative spirit and knowledge and ski list of population	• Social appropriation of scientific, technological and innovative knowledge programme. • Programme of curricular and extracurricular renewal for the promotion and development of the scientific spirit, technological, innovative and enterprising thought.
Programme of technology transfer, strategic alliances, inter-institutional articulation and technological links	• Establishment of an Innovation Centre in Bogotá programme. • Technology transfer, strategic alliances, inter-institutional articulation and technological links programme. • Incubation programme of technology-based companies to promote Creative, innovative, and entrepreneurial spirit • Programme technological development, innovation and modernization of business conglomerates. • District orientation programme on new forms of work and occupations in the city
Infrastructure for science, technology and innovation	• Programme for consolidation of innovative media for construction of the knowledge city. • Creation of the District Centre for Scientific and Technological Information and Documentation: Centre for the provision of high level Services in scientific and technological information and documentation in all areas of knowledge. • Programme to facilitate and promote access and use of ICT in support of the consolidation of the Knowledge Society.

Source: Own elaboration based on Bogotá's STI Policy

An inclination for advocacy and agreement between plans, programmes and conceptual framework to define a single line of action. This means that if the conceptual framework is focused on equity and equality, programmes and projects should also focus on those concepts. If not, a conceptual framework of development, industry and growth, as connecting axes, is recommended.

Has a clear relationship between the National Plan Development and the District Plan Development with relation to STI components. The policy has to be in agreement with the National Development Plan, of which Bogotá will be a part. In this case, "Bogotá Best for All" aims to "promote the full development of the potential of the inhabitants of the city, to achieve the happiness of all in their condition as individuals, family members and society". Politics must be focused on society, promoting access to technology and knowledge and providing tools for innovation and entrepreneurship to generate greater social welfare among citizens.

These elements are important in the formulation of urban STI policy, which concur with UNCTAD (2011) that recommend the following: (1) develop policy design and implementation through an improved urban dialogue in the STI domain; (2) improve stronger linkage and more effective interactions among interest groups of the urban/national STI systems; (3) determine measures that encourage the development of absorptive capacity by the various interest groups and facilitate the transfer of knowledge and technology through investment and channels of transfer; (4) define specific short-, medium- and long-term actions that can lead to stronger science and technological capabilities across sectors and measures to promote sectors of specific potential for technological development in the urban areas.

14.3 Methods

To develop this research and understand the main factors in the design of an adequate STI policy, using a set of guiding questions, interviews and focus groups were conducted on different interest groups and experts to allow an analysis of interrelations within the group regarding their opinions on specific topics. Focus groups provide an excellent opportunity to hear opinions, to explore issues in depth and to obtain insights that may not be heard without the discussions the focus groups provide (Palomba & Banta, 1999, p. 196). The advantages of focus groups are that the participants can express their opinions in their own language providing a multi-dimensional perspective of their knowledge, attitudes and behaviours and emphasize on specific opinions, especially when there are differences between the participants or decision-makers, and it is important to study the degree of consensus on a specific topic (Bonnett & Williams, 1998; Morgan & Kreuger, 1993). Metaplan was the method used to define different analysis categories to determine the experiences and perceptions of the respondents; it provided a rich and detailed understanding of the main factors that could affect STI policy in Bogotá.

The advantages of these techniques are as follows: (1) the participants can express their opinions in their own language to give a multi-dimensional perspective of the knowledge, attitudes and behaviours of the participants and to emphasize specific opinions (Bonnett & Williams, 1998); (2) all people who play a part in the implementation are involved, stimulating a variety of ideas that allow multiple points of view and support change initiatives from opinions of interest groups (Schnelle & Stoltz, 1987).

The group sessions and interviews were held between March and April 2017 by members of the research team with experience in these types of techniques (see Table 14.2). The discussion phase in each session ended once the subject had been exhausted by the respondents (considering every subject proposed for discussion had been exhausted). The sessions were recorded in order to extract key discussion points later from the transcriptions.

Table 14.2 Interviews and focus groups carried out within the framework of the project to determine the main factors to be considered in an STI policy

Type	*Entity/sector*
Seven Interviews	Public officials of the District Office of Planning, District Office of Economic Development, District Office Health, High information-communication technologies (ICTs) Counselling, Women Office, Education Office
Five focus groups	District functionaries (approximately 15 assistants per group)
Four focus groups	Universities and research centres (approximately 15 assistants per group)
One focus group	Productive sector (approximately 15 assistants per group)
One focus groups	Civil society (approximately 15 assistants per group)

The design of the script for the interviews and focus groups used different dimensions including the following: (1) *Motivation*: perceptions on the importance of a science, technology and innovation policy; (2) *Attitudes and values*: an analysis of the knowledge activities and relationships of science, technology and innovation; (3) *Policy instruments*: levels of information and applications of instruments of public policy regarding science, technology and innovation (funds, networks, royalties, etc.); (4) *Expectations*: the main perceptions and ideals around Bogotá as a developed city regarding science, technology and innovation; and (5) Proposals: Ideas, elements and plans to develop adequate STI policy for Bogotá.

When the interviews and the focus groups were finished, we started to analyse the principal problems that actors and groups identified, obtaining 31 factors (see Table 14.3) that are related to policy issues in science, technology and innovation. These were analysed from a matrix of cross-impacts, a tool that helps identify the problems of greater impact, that is, those that are more influential and more dependent. In the construction of this matrix, we used several qualitative techniques of analysis to build a tool of collective reflections. This implemented technique "Offers the possibility of describing a system with the help of a matrix that relates all its constituent elements" (Prospective EU, 2017). This method allows for the determination of the key factors and their relations from the perspective of the STI policy using the results of the opinions and perceptions of the different interest groups.

First, we defined the factors and gave a score according to each factor's influence on others, where "0" represents no influence, "1" represents little influence, "2" moderate influence, "3" strong influence and "P" potential influence. This matrix shows the influence of the row factors on those of the column, resulting in a Cartesian plane in which the dependence is measured on the X-axis and the influence measured on the Y-axis (Godet, 2007a).

Table 14.4 shows part of the results matrix, where it is possible to see the influence of STI investment on smart specialization. Its influence is "2", meaning that the presence of a budget for STI has a strong influence on the strategy of smart specialization, according to interviews and focus groups. Another example is the strong influence that smart specialization has on the generation of value chains, because based on interviews and focus groups, this factor has generated a greater chain of value than the public policy itself, with an influence rating of "3".

Table 14.3 List of factors obtained from interviews and focus groups

Long label	*Short label*
Investment in STI	Invest
Smart specialization	SmartSpec
Access to information	OpenInfo
Generation of value chains	ValueChain
Business training	Training
Coherence between projects and axes	Coherence
Promotion of technology development centres	DevTechCen
Promoting innovation offices inside universities	InnOffices
Promotion of inter-institutional agreements	Agreement
Academic cooperation	Cooperatio
Promote STI culture	Culture
Region focus	Regi.Focus
Entrepreneurship	Entreprene
Social focus	Soc.Focus
Policy participation in peace scenarios	Peace
Legal stability	Legal
Optimal working environments	WorkEnviro
Tax insensitive	Tax.Insens
Non-tax liabilities	NoTx Liab
Investigation-Action	Inves-Act
Maximization of benefits	Max.Benef
Work optimization	W.Optimiza
Daily problems	Day.Proble
Foster knowledge networks	NetwKnowle
Promote innovation	Innovation
University-State-Company relationship	Rel. U-S-C
Harnessing technology	Harnessi.T
Transparency	Transparen
Disruption of district policy with national science and technology policy	Disruption
Unclear objectives	Objectives
Disconnection between the conceptual framework principles and axes	Disconnect

Source: Own elaboration based on interviews and workshops

When the matrix was completed, factors and variables with names similar to those used by Godet in his prospective manual from 2007 were located inside a Cartesian plane that is distributed as shown in Table 14.5, indicating the influence of factors according to the results of interviews and focus groups.

Table 14.4 Vester Matrix

	1 : Invest	2 : SmartSpec	3 : OpenInfo
1 : Invest	0	1	0
2 : SmartSpec	2	0	0
3 : OpenInfo	0	1	0
4 : ValueChain	1	3	2
5 : Training	2	3	2
6 : Coherence	0	1	3
7 : DevTechCen	3	2	0
8 : InnOffices	0	3	0
9 : Agreement	1	2	1
10 : Cooperatio	1	2	2
11 : Culture	0	0	0

Table 14.5 Influence matrix factor typology

Factor typology		Influence ON other factors: Strong	Influence ON other factors: Weak
Influenced BY other factors	Strong	Critical	Passive
Influenced BY other factors	Weak	Active	Buffer

Source: Godet, 2007a

14.4 Results

In this section, the main results of the interviews and focus groups are analysed, taking into account the perceptions, knowledge and understanding of the importance of STI policy for Bogotá. The results are presented based on the methods used.

Once the matrix of cross-impacts has been identified, programming is performed for the Cartesian plane and the relationships between the identified factors are displayed on the Cartesian plane based on their influence (motricity) and dependence, as shown in Table 14.4. In the case of the

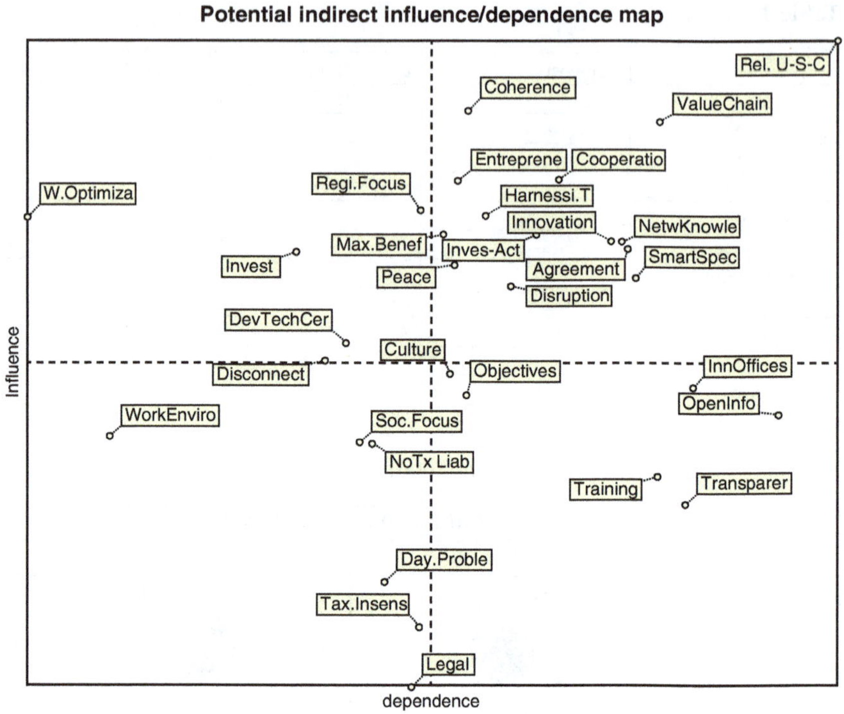

Fig. 14.1 Potential indirect influence and dependence map

STI public policy in Bogotá, the programming resulted in the Cartesian plane shown in Fig. 14.1, where the majority of the factors located on the upper part of the plane determined different relationships between influence and dependence. For example, legal stability has a low influence and average dependence, whereas work optimization has a high influence and a low dependence. The main results are listed below, according to their location on the plane.

Following Godet (2007b), the Cartesian plane shows (see Fig. 14.1) that in the upper left quadrant are the policy formation and resolution categories, and acting on them has impacts felt throughout the system, thanks to their motricity and little dependency. In this exercise, the categories located in that quadrant are work optimization, investment in STI, maximization of benefits, disconnection between the conceptual framework principles and axes, regional focus, and promotion of technology development centres.

The Cartesian plan makes it possible to identify the importance of strengthening the focus of the region, which—although it has been prevalent during the last few years—needs to be strengthened, especially in the case of smart cities, of which Bogotá seeks to be a part. This implies an increase in the investment in STI to guarantee a robust, reliable and affordable broadband network, an efficient ecosystem for the Internet of Things and the capacity to make use of the big data that are generated (United Nations, 2016).

Finally, it is necessary to have an agreement in understanding the development of the policy, because if the actions that follow are not the same, multiple interpretations can result that can hinder the implementation process and increase the problems already identified due to different entities. This concurs with United Nations (2016), which recommends the integration of approaches as an efficient tool for determining the dynamic relations between people, policies and environments. Policies are most often useful and effective when co-produced by scientists from multiple disciplines, policymakers, planners, managers, civil society representatives and all relevant interest groups.

We can see critical variables in the upper right quadrant, representing those that are closer to an average degree of dependence are similar to the active variables considered for the formulation of policy objectives, namely, policy participation on peace scenarios, foster knowledge networks and entrepreneurship.

It is important to understand that some variables with a high average dependence and motor skills can be intervening variables that do not necessarily depend excessively on other factors. However, the participation of STI policy in peace scenarios could contribute to this social and community approach that civil society and the productive sector requested in the focus groups. Encouraging and fostering knowledge networks is a fundamental element for science and the development of the city and the region. Finally, the close relationship between entrepreneurship and innovation is indisputable, as is the need for both, in the city, to contribute to competitiveness, productivity and optimization. Multi-disciplinary and holistic STI is fundamental because it can help achieve the Sustainable Development Goals (SDGs) and improve people's lives using different sources of knowledge that are translated into meaningful improvements in day-to-day lives and economic growth and development (SDGs, 2017).

The following variables are also critical and are considered the strongest variables in terms of movement and dependence on STI issues in the city and other factors: disruption of district policy with national science and technology policy, disconnection between the conceptual framework principles and axes, University-State-Company relationship, generation of value chains, academic cooperation, investigation-action, harnessing technology, promote innovation, promotion of inter-institutional agreements and smart specialization. These factors could be affected as well, similar to factors mentioned above, and demonstrate the importance of including key factors to achieve an adequate and coherent STI policy that allows for improving efficiency in both economic and environmental terms, building new and more sustainable alternatives to satisfy human needs and empowering citizens to drive their own future (European Commission, 2015).

The variables that we mentioned earlier are the strongest and most important for STI in Bogotá, but they also have a large dependence on other variables and therefore can benefit from the influence of these variables. Indifferent variables, those found in the lower left quadrant, are identified as follows: work optimization, social focus, daily problems, non-tax liabilities, tax insensitive and legal stability. These variables can be monitored based on the results of the STI policy implemented in the city to ensure that planned progress is made and the desired results achieved (Brito, 2014).

Although the results indicate that for different entities these factors are important within the STI policy, there are factors that do not strongly affect each other, but when other factors intervene, they can benefit. For example, when training is involved in companies, it in turn can have an impact on an optimal work environment, but an optimal work environment does not necessarily mean one that includes business training. In addition, there are other variables that do not necessarily intervene in the system.

Finally, there are variables located in the lower right quadrant, which are the result variables or buffers. They are variables that have a greater dependence and less influence, and once they are activated, they are going to be the first ones affected and will respond in the system, such as promote STI culture, objectives, promoting innovation offices inside universities, access to information, and business training and transparency. It is fundamental to continue with the promotion of open governments and citizen control to help make better decisions and enable a more active society, creating value through transparency, democratic control, self-empowerment, improved or new private products and services, improved efficiency and effectiveness of government services, impact measurement of policies and new knowledge from combined data sources and patterns in large data volumes.

Both the interest groups and the experts show the need to increase investment in science, technology and innovation in Bogotá, ensuring a decrease in many problems and promoting scientific collaboration to develop new knowledge, new technologies and innovation to apply in all productive sectors. This will stimulate and maintain economic growth and sustainable development, and is consistent with studies by McDonald and Bailly (2017) in the urban context and Hammer (2011) in studies of green projects and research and development.

Moreover, it is necessary to formulate programmes and projects that have implications for society and aim to optimize work, both in the State, universities and enterprises, which could be carried out through the creation and development of technological development centres (see Fig. 14.2). This type of cooperation produces a positive effect on innovation, productivity and economic growth performance and is reinforced when the enterprise has a high-growth orientation, especially in developing countries (Audretsch, 2012; Autio, Kenney, Mustar, Siegel, & Wright, 2014; Guerrero & Urbano, 2017).

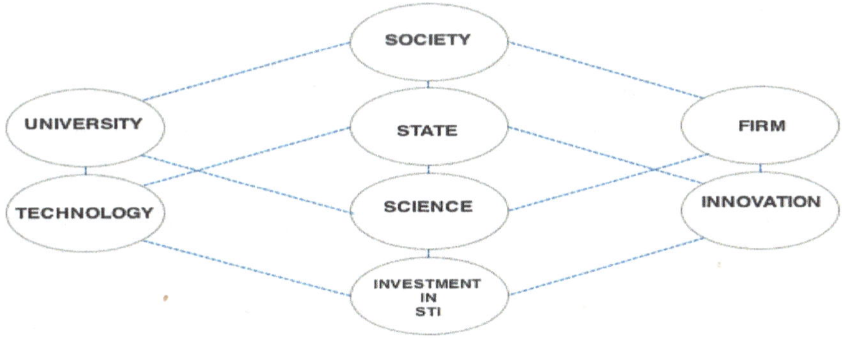

Fig. 14.2 Relational analysis for role the State, universities and enterprise

14.5 Discussion

Based on an analysis of factors, a matrix of cross-impacts and a revision of the District's science, technology and innovation policy, the following action plan for the formulation of the new STI policy is observed and recommended.

The objective of the policy is to be a state policy and not a government policy, meaning that it lasts over time and there is no change of policy every four years, as occurs in the case of Colombia, because it is in this period that the policy begins to generate impact. However, periodic monitoring should be carried out to determine if a reformulation or modification is required, especially for science, innovation and technology due to the rapid STI changes that are taking place in the world.

It is important to take into account the rationality of the policy and important to consider a participatory component that is referred to by Torgerson (1986) as the communicative rationality, which includes the importance of the technical and political component. The first component is related to the formulation and scope of public policy and the second to the importance of the issue on the political agenda, because if there are no such interests, the implementation of different plans and projects will be limited.

It is important because of the humanistic focus that public policies have had in Bogotá. The active participation of citizens is fundamental in the policy formulation, and due to the different priorities and problems that the social groups have, the general objectives need to be clearly defined.

Once defined, the general objectives should be in accordance with the National STI Policy: "To increase the capacity of the country to identify, produce, disseminate, use and integrate scientific and technological knowledge, with the purpose of improving competitiveness, and contribute to the productive transformation of the country". Based on the material collected, specific objectives can be defined as follows:

- Increase investment in science, technology and innovation in Bogotá to increase scientific production, transfer of knowledge and use of technologies.
- Develop technological development centres that encourage companies to optimize work and maximize profits.
- Be recognized as a policy that has driven the region to have greater economic and social development.

- Through science, technology and innovation as transversal axes in the different sectors, promote projects and programmes that contribute to peace scenarios in the region.
- Encourage innovation through an increase in the number of new companies that generate greater development in Bogotá and the region.

Therefore, it is recommended to use the value chain method, which is more advanced than the logical framework methodology and is currently used more by policymakers, by including key additional elements such as inputs, activities, products and results. Figure 14.3 shows an example.

The advantage of this methodology is that could respond to more than one objective, so it is evident that the whole chain works in favour of a single and a general objective, and therefore, an intervention of many factors is evident, as was mentioned in the analysis of the model presented in the previous section.

It is also essential to consider collective governance in science, technology and innovation, since modern public policies do not fall strictly in the public sector but are a joint effort of the public sector, the private sector, the third sector and citizens. Therefore, it is important to publicize the policy, socialize the policy, and provide open information and public accountability. For although it is the State that must formulate the policy, its implementation and success will depend on joint work. This does not

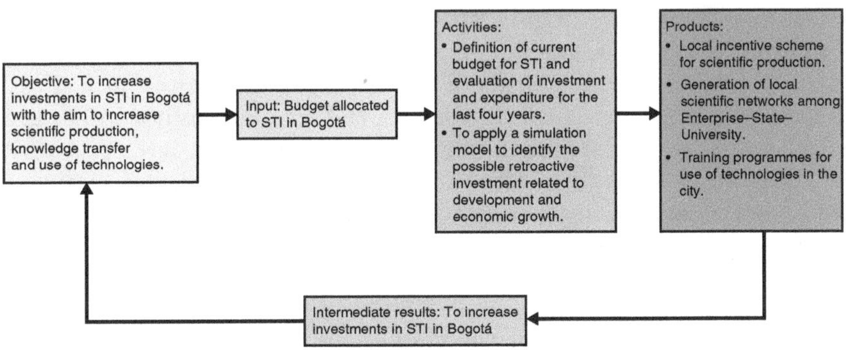

Fig. 14.3 Value chain in the formulation of STI policy

mean that the scientific committee should be composed of a representative of each entity or institution but that STI is an issue that must involve all citizens.

It is necessary to clarify the terminology that will be used within the public policy, since from the interviews, focus groups and conversations, it is evident that a single concept of science, technology and innovation does not exist, and different entities are adopting their own concept and identifying problems and needs from it.

Finally, it is important to formulate a measurement system for STI in the District, and it is necessary to constantly monitor the different programmes to identify faults in a timely manner and intervene if necessary. In addition to this, it is important to publish this information in different media, enabling the citizens, the private sector and the third sector to participate in the development of the policy.

14.6 Conclusions

In general, this study shows that the participants of the interviews and focus groups believe that it is important to improve the STI policy and concur with the different factors that should be included in the new policy, so as to increase investments in STI, to focus the concept on the Bogotá region and to maximize benefits from STI, indicating the importance of analysing policies from the perspectives of different interest groups to achieve an adequate STI policy. In addition, techniques and models that allow for categorization were used and helped to determine inputs for the development of effective policy.

To reformulate the Bogotá STI policy, it is necessary to consider that more projects and programmes do not ensure a more efficient and successful policy. In fact, the existence of many plans and projects generates a dependency on other plans and projects, and when one is not working, a vicious circle is generated that does not allow progress in development. It is advisable to focus on projects and plans of those variables that were active in this analysis since these factors, due to the influence they have on other factors, would cause change and improvement, thus eliminating, in the long term, those critical factors that have been identified as major problems in relation to STI.

An STI policy must be open and not exclusive, as it is required in other fields, such as education, health, finance and the environment to improve competitiveness, optimization, and the human capital and development of the region. This does not mean that each sector must have its own science, technology and innovation programme, but this development must be achieved so that the presence of these components is permanently present in all sectors.

Investment in STI is necessary in Bogotá and the region, in addition to involving citizens more in discussing the subject of STI, demonstrating that it is not just the use of a smart phone or a computer. In addition, it is important to rely on the tool, smart specialization, to strengthen the regional efforts to reach the goal of becoming a smart city. Innovation must play a fundamental role, both for the entrepreneur and the businessman, because thanks to this tool, benefits can be maximized and work optimized. Finally, we cannot ignore the necessary role of the State as a facilitator and maker of politics and not as an impediment to the company and university to intervene and advance the issues of STI.

References

Audretsch, D. (2012). *Determinants of High-Growth Entrepreneurship Report Prepared for the OECD/DBA International Workshop on High-Growth Firms: Local Policies and Local Determinants.* Copenhagen, 28 March 2012.

Autio, E., Kenney, M., Mustar, P., Siegel, D., & Wright, M. (2014). Entrepreneurial Innovation: The Importance of Context. *Research Policy, 43*(7), 1097–1108.

Bonnett, M., & Williams, J. (1998). Environmental Education and Primary Children's Attitudes Towards Nature and the Environment. *Cambridge Journal of Education, 28*, 159–174.

Brito, L. (2014). The Role of Science, Technology and Innovation Policies and Instruments for a Paradigm Shift Towards Sustainable Development. Chapter 2 of Book *Technologies for Sustainable Development.* https://doi.org/10.1007/978-3-319-00639-0_2

Commission on Science and Technology for Development. (2013). *Science, Technology and Innovation for Sustainable.* Economic and Social Council, 1–19.

Conpes. (2009). *National Policy of Science, Technology and Innovation.* Bogotá, D.C.

District Commission for Science, Technology and Innovation. (2007). *Science, Technology and innovation Plan Bogotá. 2007–2019.* Bogotá.

European Commission. (2015). Quality of Life in European Cities. https://ec.europa.eu/regional_policy/sources/docgener/studies/pdf/urban/survey2015_en.pdf

Godet, M. (2007a). *Tools Box of Strategic Prospective.* Paris: LIPS Books.

Godet, M. (2007b). *Tools Box of Strategic Prospective.* Paris: Cuadernos de LIPS.

Guttieres, D., Fransiska, G. A., & Kusuma Dewi, S. R. (2016). *Role of Science, Technology & Innovation in Urban Frameworks: Enhancing the Science-Policy-Practice Interface for Resilient Cities.* Brief for GSDR.

Guerrero, M., & Urbano, D. (2017). The Impact of Triple Helix Agents on Entrepreneurial Innovations' Performance: An Inside Look at Enterprises Located in an Emerging Economy. *Technological Forecasting and Social Change, 119*, 294–309.

Hammer, S. (2011). *Cities and Green Growth: A Conceptual Framework.* OECD Regional Development Working Papers 2011/08, OECD Publishing. https://doi.org/10.1787/5kg0tflmzx34-en

McDonald, R., & Bailly, A. (2017). What Investors Want: A Guide for Cities. https://www.centreforcities.org/publication/investors-want-guide-cities/

McFarlane, C., & Söderström, O. (2017). On Alternative Smart Cities: From a Technology-Intensive to a Knowledge-Intensive Smart Urbanism. *City, 21*(3–4), 312–328.

Morgan, D., & Krueger, R. (1993). When to Use Focus Groups and Why. In D. Morgan (Ed.), *Successful Focus Groups: Advancing the State of the Art* (pp. 3–20). Newbury Park, CA: Sage.

National Council of Innovation for Competitiveness. (2012). *National Strategy of Innovation*. Santiago.

OECD. (2012). *National Strategies for Science, Technology and Innovation*. STI Policy Profiles: Innovation Policy Governance, 146–148.

Palomba, C. A., & Banta, T. W. (1999). *Assessment Essentials: Planning, Implementing, and Improving Assessment in Higher Education*. San Francisco, CA: Jossey-Bass.

Prospective EU. (2017). *MIC MAC. Structural Analysis.*

Schnelle, W., & Stoltz, I. (1987). *The Metaplan Method: Communication Tools for Planning Learning Groups* (Metaplan Series No. 7). Goethestrasse, Germany.

Science, Technology and Innovation Ministry. (2015). *National Plant of Science, Technology and Innovation (STI) 2015–2021.* San Jóse.

Sela. (2016a). *Science, Technology and Innovation in Latin America & Caribe Public Policy.* Lima: Ecuador Case.

Sela. (2016b). *Public Policies on Science, Technology and Innovation in Latin America and Caribe.* Lima: Ecuador Case.

Sustainable Development Goals (SDGs). (2017). *UN Moves to Tap Transformative Power of Science, Technology and Innovation to Achieve Sustainable Development Goals.* Retrieved from http://www.un.org/sustainabledevelopment/blog/2017/05/un-moves-to-tap-transformative-power-of-science-technology-and-innovation-to-achieve-sustainable-development-goals/

Torgerson. (1986). Between Knowledge and Politics: Three Faces of Policy Analysis. *Policy Science, 19*, 33–59.

UNCTAD. (2011). *A Framework for Science, Technology and Innovation Policy Reviews.* Retrieved from http://unctad.org/en/Docs/dtlstict2011d7_en.pdf

United Nations. (2016). *Smart Cities and Infrastructure.* Report of the Secretary-General. Retrieved from http://unctad.org/meetings/en/Sessional Documents/ecn162016d2_en.pdf

Glossary of Acronyms

ATECMA Aeronautical Sectoral Association
BAN British Approved Name
BDCOL Biblioteca Digital Colombiana (in English: Colombian Digital Library)
CAN Community of Andean Nations
CBD Convention on Biological Diversity
CEAACES Committee of Evaluation, Accreditation and Higher Quality Assurance
CONACYT National Council of Science and Technology
CONESUP Consejo de Educación Superior
CONPES National Council of Political and Economic Policies
CPI Consumer price index
CRA Potable Water and Basic Sanitation Regulatory Commission
DANE National Administrative Department of Statistics
DNP National Planning Department
ECLAC Economic Commission for Latin America and the Caribbean
ENSO El Niño–Southern Oscillation
EPO European Patent Office
EU European Union
FDA Food and Drug Administration
GDP Gross domestic product
GII Global Innovation Index

C. I. Pardo Martínez et al. (eds.), *Analysis of Science, Technology, and Innovation in Emerging Economies*,
https://doi.org/10.1007/978-3-030-13578-2

HD Historical-documentary
HDI Human Development Index
ICS Supervisory Agency of Industry and Commerce
ICT Information and Communication Technologies
IDEAM Institute of Hydrology, Meteorology and Environmental Studies
IES Institutions of Higher Education
INSEAD European Institute of Business Administration
IPC International Patent Classification
JPO Japan Patent Office
LAC Latin American and the Caribbean
LOES Organic Law of Higher Education
MCA Multiple Correspondence Analysis
MINSALUD Ministry of Health and Social Protection
MRO Maintenance, repair, and overhaul
MTI Means of technological innovation
NCEs New chemical entities
NGOs Non-governmental organizations
NTDs New therapeutic drugs
OA Open Access
OCyT Colombian Observatory of Science and Technology
OECD Organization for Economic Cooperation and Development
PCT Patent Cooperation Treaty
PISA Program for International Student Assessment
PS&T Scientific technological policy
REMCI Ecuadorian Network of Female Scientists
RENATA National Network of Research and Education
R&D Research and development
RICYT Network for Science and Technology Indicators–Ibero-American and Inter-American
SAS&T Social appropriation of science and technology
SC Scientific sub-competences
SDGs Sustainable Development Goals
SENESCYT Secretaría de Educación Superior, Ciencia, Tecnología e Innovación
SIC Superintendence of Industry and Commerce
SMEs Small and medium enterprises
STEM Science, Technology, Engineering, and Mathematics
STI Science, technology and innovation
STIA Scientific, technological and innovation activities

STS Science, Technology, and Society
SPPS&T Surveys on public perception of science and technology
SWOT Strengths, weaknesses, opportunities, and threats
TFP Total factor of productivity
TIS Technological innovation systems
UNASUR Unión de Naciones Suramericanas
UNDP United Nations Development Program
UNESCO United Nations Educational, Scientific and Cultural Organization
USAN United States Adopted Name
USPTO United States Patent and Trademark Office
WB World Bank
WIPO World Intellectual Property Organization

FSC
www.fsc.org
MIX
Papier | Fördert gute Waldnutzung
FSC® C083411

Kolibri 360 GmbH
Ferdinand-Jühlke-Straße 7
99095 Erfurt, Deutschland
produktsicherheit@kolibri360.de